# STANDARD GRADE GEOGRAPHY

# *International*

# *Issues*

## Third Edition

A dirt road running through the rainforest in West Kalimantan, Borneo, providing access for loggers who are clearing the forest to create new settlements.

## CALVIN CLARKE

Hodder Gibson
A MEMBER OF THE HODDER HEADLINE GROUP

The publishers would like to thank the following for permission to reproduce copyright material:

**Photo credits**
Page 14, © Hutchinson Library/Brooke Hyde; page 19, © Life File/ Richard Powers; page 20, © Life File/ Andrew Ward; page 21, © Jan Butchofsky-Houser/Corbis; page 28, © Hutchinson Library/Liba Taylor; page 30, © Hutchinson Library/Liba Taylor; page 36, © Achinto/Christian Aid/Still Pictures; page 46, © Paul Harrison/Still Pictures; page 55, © Yann Arthus-Bertrand/Corbis; page 60, © Jérome Sessini/In Visu/ Corbis; page 64, © Reuters/Corbis; page 80, © Jacques Jangoux/Still Pictures; page 93, © Life File/Louise Oldroyd; page 100, © Sean Sprague/Still Pictures; page 101, © Mark Edwards/Still Pictures; page 108, © The Fairtrade Foundation; page 118, © Glen Christian/Still Pictures; page 120, © Hutchinson Library/ Anna Tully; page 125, © Hugh Rooney/Eye Obiquitous

Every effort has been made to trace all copyright holders, but if any have been inadvertently overlooked the Publishers will be pleased to make the necessary arrangements at the first opportunity.

Although every effort has been made to ensure that website addresses are correct at time of going to press, Hodder Gibson cannot be held responsible for the content of any website mentioned in this book. It is sometimes possible to find a relocated web page by typing in the addressof the home page for a website in the URL window of your browser.

Orders: please contact Bookpoint Ltd, 130 Milton Park, Abingdon, Oxon OX14 4SB. Telephone: (44) 01235 827720. Fax: (44) 01235 400454. Lines are open from 9.00 – 5.00, Monday to Saturday, with a 24-hour message answering service. Visit our website at www.hoddereducation.co.uk. Hodder Gibson can be contacted direct on: Tel: 0141 848 1609; Fax: 0141 889 6315; email: hoddergibson@hodder.co.uk

© Calvin Clarke 2005
First published in 2005 by
Hodder Gibson, a member of the Hodder Headline Group, an Hachette Livre UK Company
2a Christie Street
Paisley PA1 1NB

Impression number   10 9 8 7 6 5 4 3 2
Year                          2010 2009 2008 2007

Cover photo © Wayne Lawler; Ecoscene/Corbis
Typeset in Stone Serif 10.5/13.5pt by Fakenham Photosetting Limited, Fakenham, Norfolk.
Printed in Dubai

A catalogue record for this title is available from British Library.

ISBN-13: 978-0-340-88926-8

# Contents

# Introduction

## Structure

This book is designed for pupils of all abilities to learn key ideas 6 and 12–17 of the Scottish Standard Grade Geography syllabus and to develop the gathering and processing techniques prescribed in the syllabus. The book is divided into 15 units and each unit is divided into several sections:

1 All pupils read the **Core text**.
2 All pupils answer the **Core questions**.

### Assessable elements

The Core questions and Extension questions in each unit test mostly the **knowledge and understanding** assessable element. The Foundation, General and Credit questions test mostly the **enquiry skills** assessable element.

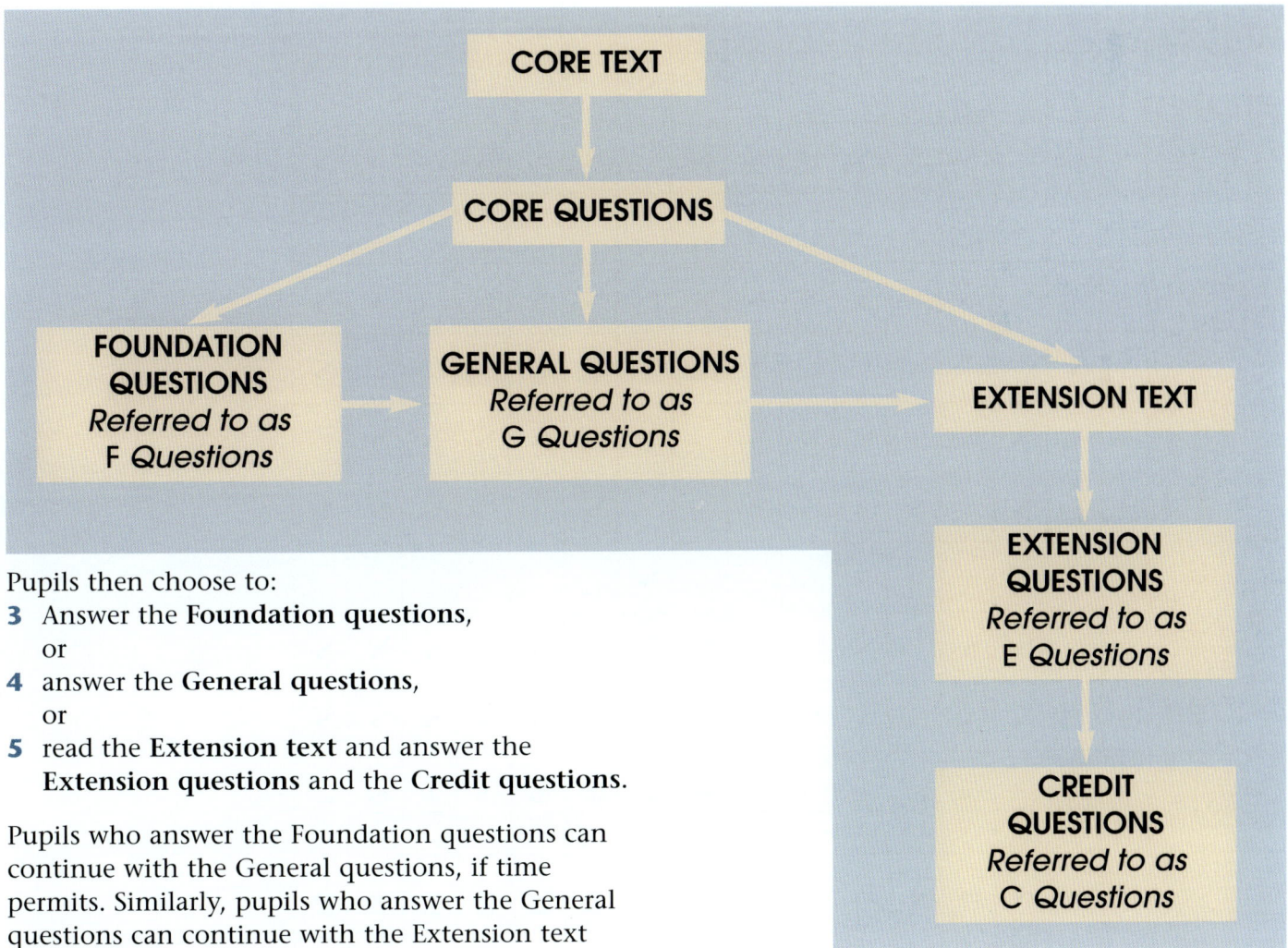

**CORE TEXT**

**CORE QUESTIONS**

**FOUNDATION QUESTIONS**
*Referred to as
F Questions*

**GENERAL QUESTIONS**
*Referred to as
G Questions*

**EXTENSION TEXT**

**EXTENSION QUESTIONS**
*Referred to as
E Questions*

**CREDIT QUESTIONS**
*Referred to as
C Questions*

Pupils then choose to:
3 Answer the **Foundation questions**, or
4 answer the **General questions**, or
5 read the **Extension text** and answer the **Extension questions** and the **Credit questions**.

Pupils who answer the Foundation questions can continue with the General questions, if time permits. Similarly, pupils who answer the General questions can continue with the Extension text and questions. Only the most able pupils should go straight from the Core questions to the Extension text.

| UNIT | KEY IDEAS | KNOWLEDGE AND UNDERSTANDING (F/G LEVELS) | | | KNOWLEDGE AND UNDERSTANDING (G/C LEVELS) | | | ENQUIRY SKILLS (F LEVEL) | | | | | ENQUIRY SKILLS (G LEVEL) | | | | | ENQUIRY SKILLS (C LEVEL) | | | | |
|---|---|---|---|---|---|---|---|---|---|---|---|---|---|---|---|---|---|---|---|---|---|---|
| | | a | b | c | a | b | c | a | b | c | d | e | a | b | c | d | e | a | b | c | d | e |
| 1 | 12 | | | ✓ | | | ✓ | | | | ✓ | ✓ | | | | ✓ | ✓ | | | | ✓ | ✓ |
| 2 | 13 | ✓ | ✓ | | ✓ | ✓ | | ✓ | ✓ | | | | ✓ | | | | | ✓ | | ✓ | | |
| 3 | 12 | ✓ | ✓ | | ✓ | ✓ | | ✓ | ✓ | ✓ | | | ✓ | ✓ | ✓ | | | ✓ | ✓ | ✓ | | |
| 4 | 13 | ✓ | | | ✓ | ✓ | ✓ | ✓ | ✓ | ✓ | | | ✓ | ✓ | ✓ | | | ✓ | ✓ | ✓ | | |
| 5 | 14 | ✓ | ✓ | | ✓ | ✓ | | ✓ | ✓ | ✓ | | | ✓ | ✓ | | | | ✓ | ✓ | ✓ | | |
| 6 | 14 | ✓ | ✓ | | ✓ | ✓ | | ✓ | | ✓ | | | ✓ | ✓ | ✓ | | | ✓ | ✓ | ✓ | | |
| 7 | 14 | ✓ | ✓ | | ✓ | ✓ | | ✓ | ✓ | ✓ | | | ✓ | ✓ | ✓ | | | ✓ | ✓ | ✓ | | |
| 8 | 14 | | | | | | | | | ✓ | | | | | ✓ | | | | | | | |
| 9 | 6 | | | ✓ | | | ✓ | | | | ✓ | ✓ | | | | ✓ | ✓ | | | | ✓ | ✓ |
| 10 | 6 | ✓ | ✓ | | ✓ | ✓ | | ✓ | ✓ | ✓ | | | ✓ | ✓ | ✓ | | | ✓ | ✓ | ✓ | | |
| 11 | 6 | ✓ | ✓ | | ✓ | ✓ | | ✓ | ✓ | ✓ | | | ✓ | ✓ | | | | ✓ | ✓ | ✓ | | |
| 12 | 6 | ✓ | ✓ | | ✓ | ✓ | | ✓ | ✓ | ✓ | | | ✓ | ✓ | ✓ | | | ✓ | ✓ | ✓ | | |
| 13 | 16 | ✓ | ✓ | | ✓ | ✓ | | ✓ | ✓ | ✓ | | | ✓ | ✓ | ✓ | | | ✓ | ✓ | ✓ | | |
| 14 | 17 | ✓ | ✓ | | ✓ | ✓ | | ✓ | ✓ | ✓ | | | ✓ | ✓ | ✓ | | | ✓ | ✓ | ✓ | | |
| 15 | 15 | ✓ | ✓ | | ✓ | | ✓ | ✓ | | ✓ | | | ✓ | ✓ | | | | ✓ | ✓ | ✓ | | |

# 1 Skills in Population Studies

## Core text

### 1A Introduction to population studies

Geography studies landscapes, both physical and human landscapes. One of the most important parts of any landscape in the world is its people, in particular where they live and what they do. Where people live and what they do depends very much on the physical landscape and what people do directly affects the human landscape. Population studies, therefore, is central to Geography.

For the Standard Grade examination, you need to know and understand the following:

1  Methods of finding out the number of people in a country.
2  Reasons why some areas are crowded and others are empty.
3  Ways of measuring the standard of living of people in a country.
4  Reasons why some areas have a higher standard of living than others.
5  Reasons why some areas have different birth rates and death rates than others.
6  Effects on countries growing in population very rapidly and very slowly.
7  Reasons why people migrate within countries and between countries.

You also need to develop the following enquiry skills:

1  How to gather information on local population by undertaking questionnaires.
2  How to process the information obtained by drawing line graphs, scattergraphs, population pyramids and dot distribution maps.
3  How to analyse the information.

The rest of this unit deals with ways of gathering and processing information about populations.

### 1B Gathering information

There are several ways of finding out information about population. They are called **gathering techniques**. To gather population information about countries, you need to use official statistics. But to investigate population in your local area, these are some of the techniques that can be used:

| TOPIC STUDIED | GATHERING TECHNIQUE |
| --- | --- |
| Population characteristics Population distribution | using a questionnaire observing and recording population on a map, extracting information from a distribution map |
| Birth rates and population growth Migration | interviewing people interviewing people |

### 1C Compiling and using a questionnaire

A **questionnaire** is a list of questions you ask to a number of people. Before you use a questionnaire, you need to answer the following questions:

- What type of questionnaire is best? Should I go from house to house asking questions, stand in the street or post the questionnaire to peoples' houses?

- When should I conduct the questionnaire? Are some days of the week better than others? Are some times of day more suitable than others?

- How many questions should I ask? Only ask the questions you really need because, generally, people do not like answering questions.

Begin by explaining why you are asking questions.

- How many people should I ask?
  Ask at least 30 people at random.

- What questions should I ask?

- This depends on what you want to find out. If you wanted to compare the populations of two areas in a town, you might wish to ask the questions listed in Figure 1.1.

Date: _____ Time: _____ Location: _____

Excuse me, I'm doing some work for my Geography course at school.
Would you mind answering a few questions?

(1) How many people are there in your household?
(2) What are their ages? Are they male or female?
(3) For how many years have you lived here?
(4) What types of job do the people in your household have?
(5) At what ages did you all leave full time education?
(6) How many (a) cars (b) TV sets are there in your household?

| QUESTION | PERSON 1 | PERSON 2 | PERSON 3 | PERSON 4 | PERSON 5 |
|---|---|---|---|---|---|
| 1 | | | | | |
| 2 | | | | | |
| 3 | | | | | |
| 4 | | | | | |
| 5 | | | | | |
| 6 | | | | | |

**FIGURE 1.1**

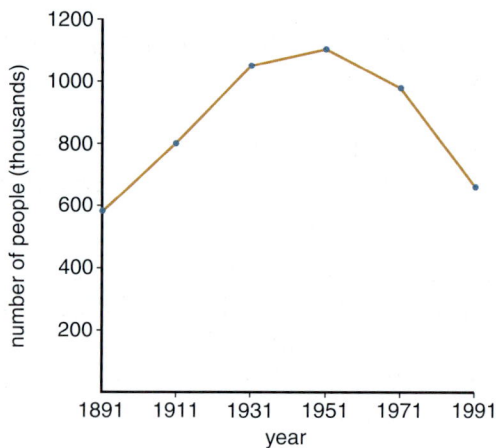

**FIGURE 1.2** *Line graph showing the population of Glasgow (1891–1991)*

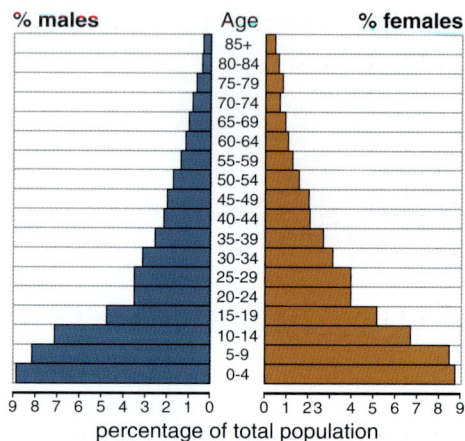

**FIGURE 1.3** *Population Pyramid for Bangladesh*

## 1D Processing techniques

Once you have gathered the information you need, your findings need to be studied and analysed. To do this, you need to organise your findings so they are easy to understand and clear to use. For example, much of the information you find out can be made into tables, maps, graphs or diagrams. These are much clearer ways of showing information. When you change your raw findings into a more useful form, you are using **processing techniques**. Some useful processing techniques in population studies are now described in detail.

## 1E Drawing a line graph

A line graph is used to show how one amount changes over time or distance.

- You should plot time or distance on the horizontal (*x*) axis, for example in Figure 1.2 the *x* axis shows **years**.

- The *y* axis should show **amounts**, for example, in Figure 1.2 it shows the **number** of people.

- Draw the axes in pencil.

- Find the highest and lowest values on the x and y axes and then choose a suitable scale for each axis.

- Label the axes in ink, including the units.

- Plot each point carefully, in pencil, with a small cross.

- Join the crosses with a pencil line.

- Once you have checked your graph, go over the pencil lines in ink.

- Write a title for your graph.

## 1F Drawing a population pyramid

A population pyramid is a type of bar graph. It is used to show the number of males and females of different age groups living in an area.

- Draw an *x* axis and, from the middle of it, draw upwards two vertical lines about one centimetre apart (as in Figure 1.3).

- The *x* axis shows the population of the area, either the total number of people or as a percentage of the total population.

- The number of males are shown on the left and the females on the right.

- Find the highest numbers on the *x* axis and work out a suitable scale.

- The wide *y* axis shows age groups, usually in five year groupings, with the youngest age group at the bottom (as in Figure 1.3).

- Draw, in pencil, a bar to show the number of males in the youngest age group.

- Then draw a bar for the females in the youngest age group.

- Continue by drawing in the bars for the next youngest age group, and so on.

- When you have finished and checked your graph, go over the lines in ink.

- Label the axes clearly.

- Give the population pyramid a title.

## 1G Drawing a scattergraph

A scattergraph is used to find out if there is a connection (or relationship) between two sets of figures.

- Draw an *x* and a *y* axis.

- Decide which information should be shown on each axis, for example in Figure 1.4, the *x* axis shows population per doctor and the *y* axis shows GNP per person.

- If one of the amounts is a distance or time, this should be shown on the *x* axis.

- Find out the highest numbers on the x and y axes and work out a suitable scale for each axis.

- Draw the axes in pencil first.

- Plot each point carefully with a small cross, but do **not** join up the crosses.

- When you have finished and have checked your graph, go over all the lines and crosses in ink.

- Label the axes clearly.

- Give the graph a title.

- Once the scattergraph is complete, you should be able to observe the pattern and identify any trends, for example in Figure 1.4, the graph shows that, generally, as GNP per person increases, the population per doctor decreases.

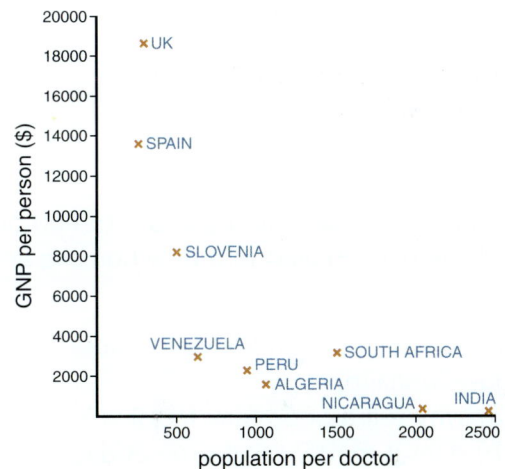

**FIGURE 1.4** *Scattergraph showing the relationship between GNP/person and population/doctor*

## 1H Drawing a dot distribution map

A dot distribution map shows the spread or pattern of any feature over an area (for example people, farms, factories).

- Your area of study should be divided into regions and you should have information for each region, for example in Figure 1.5 northern Scotland has been divided into seven local Government areas.

- Decide the value of each dot, making sure there are enough dots to show the distribution accurately, but not too many so that the map is overcrowded, for example in Figure 1.5, each dot represents 30 000 people.

- Decide on the size of the dots, making sure they are all the same size.

- Work out how many dots each region should have and then plot the dots on the map, in pencil first.

- The dots should be plotted evenly over each region, but should also be placed over known locations, for example in Figure 1.5, the Highland district has seven dots and two of these have been placed directly over the two biggest towns (Fort William and Inverness) and the others placed in the areas where most people live.

- When you have finished, go over the dots in coloured pencil or ink.

- Give the map a title and a key.

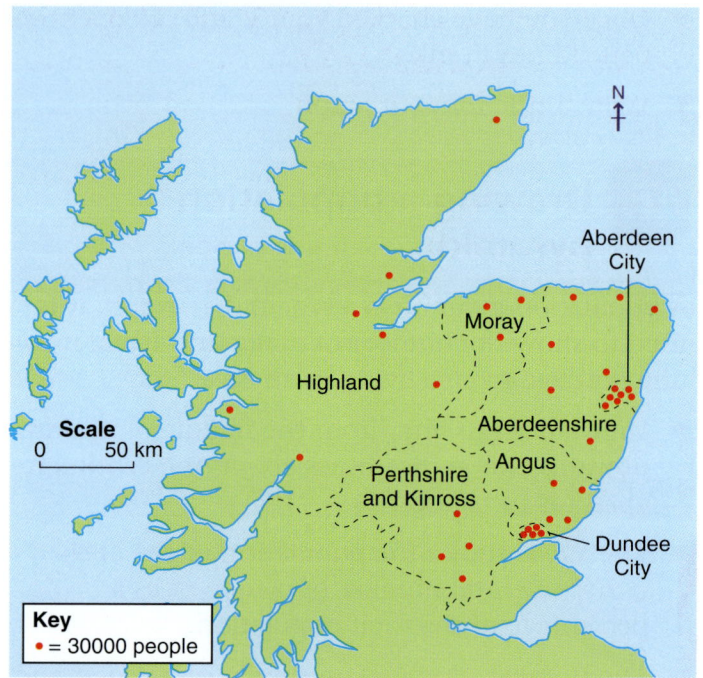

**FIGURE 1.5** *Dot distribution map showing the population distribution in the northern mainland of Scotland*

## FOUNDATION QUESTIONS

**1** *Look at 1B.*
What technique would you use to find out why people have moved to a new housing area?

**2** *Look at 1C.*
If you were doing a house-to-house questionnaire,
(a) why is it a good idea to ask a lot of people?
(b) does it matter what time of day you do the questionnaire?

**3** *Look at 1E.*
(a) Draw the line graph shown in Figure 1.6.
(b) Complete the graph using the information in Figure 1.7.

| YEAR | POPULATION OF EAST KILBRIDE |
|------|------------------------------|
| 1951 | 6000 |
| 1956 | 15 000 |
| 1961 | 33 000 |
| 1966 | 49 000 |
| 1971 | 65 000 |
| 1976 | 71 000 |
| 1981 | 73 000 |
| 1986 | 77 000 |
| 1991 | 69 000 |
| 1996 | 72 000 |
| 2001 | 74 000 |

**FIGURE 1.7**

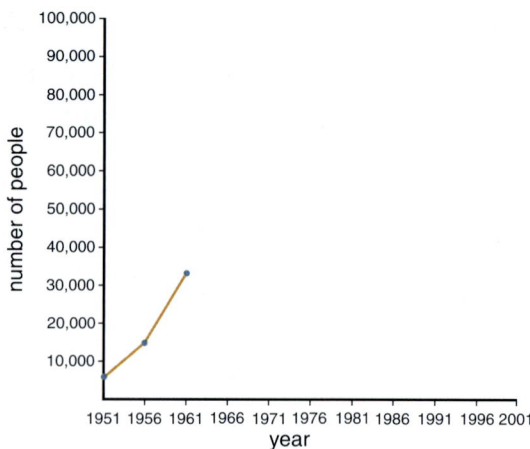

**FIGURE 1.6**

**4** *Look at 1G.*
  (a) Draw the scattergraph shown in Figure 1.8.
  (b) Complete the scattergraph using the information in Figure 1.9 below.

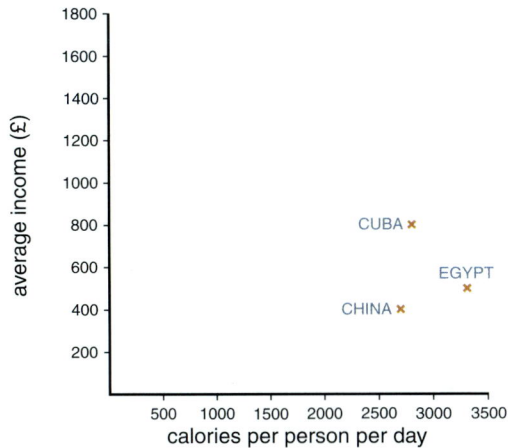

**FIGURE 1.8**

| COUNTRY | AVERAGE INCOME (£) | CALORIES PER PERSON PER DAY |
|---------|--------------------|-----------------------------|
| China | 400 | 2700 |
| Cuba | 800 | 2800 |
| Egypt | 500 | 3300 |
| Fiji | 1500 | 3100 |
| India | 200 | 2400 |
| Iraq | 1100 | 2100 |
| Nigeria | 200 | 2100 |
| Turkey | 1700 | 3400 |

**FIGURE 1.9**

**5** *Look at Figure 1.10.*
  Which table of information (A, B or C) could be shown by (a) a population pyramid, (b) a dot distribution map?

| EUROPEAN UNION COUNTRY | POPULATION (THOUSANDS) |
|------------------------|------------------------|
| Austria | 8200 |
| Belgium | 10 300 |
| Denmark | 5400 |
| Finland | 5200 |
| France | 59 800 |
| Germany | 83 300 |
| Greece | 10 600 |
| Ireland | 3900 |
| Italy | 57 700 |
| Luxembourg | 400 |
| Netherlands | 16 100 |
| Portugal | 10 100 |
| Spain | 40 100 |
| Sweden | 8900 |
| United Kingdom | 59 800 |

**FIGURE 1.10A**

| AGE GROUP | GERMANY MALE POPULATION (%) | FEMALE POPULATION (%) |
|-----------|-----------------------------|------------------------|
| 0–9 | 5.5 | 5.0 |
| 10–19 | 5.5 | 5.5 |
| 20–29 | 7.5 | 7.5 |
| 30–39 | 7.5 | 7.5 |
| 40–49 | 6.0 | 6.0 |
| 50–59 | 7.0 | 7.0 |
| 60–69 | 4.5 | 5.0 |
| 70–79 | 3.0 | 4.0 |
| 80+ | 2.5 | 3.5 |

**FIGURE 1.10B**

## FOUNDATION QUESTIONS

| YEAR | CHINA'S POPULATION DENSITY |
|---|---|
| 1950 | 59 people per sq. km |
| 1960 | 71 |
| 1970 | 89 |
| 1980 | 107 |
| 1990 | 124 |
| 2000 | 138 |

**FIGURE 1.10C**

| COUNTRY | BIRTH RATE (PER 1000 PEOPLE) | INFANT MORTALITY (PER 1000 BIRTHS) |
|---|---|---|
| Australia | 14 | 5 |
| Brazil | 20 | 53 |
| Cambodia | 43 | 106 |
| Congo | 48 | 106 |
| Ethiopia | 46 | 122 |
| Indonesia | 23 | 61 |
| Japan | 10 | 4 |
| South Africa | 27 | 53 |

**FIGURE 1.10D**

## GENERAL QUESTIONS

1. ***Look at 1B.***
   What technique would you use to find out the population density of a housing area? Give a reason for your answer.

2. ***Look at 1C.***
   If you were carrying out a questionnaire to find out how long people have lived in an area,
   (a) would you go from house to house asking questions or stand outside the local shops? Give a reason for your answer.
   (b) Describe two other points you should consider before carrying out your questionnaire.

| COUNTRY | LIFE EXPECTANCY | POPULATION PER DOCTOR |
|---|---|---|
| Brazil | 62 | 1000 |
| Guinea | 46 | 7700 |
| Iran | 68 | 3300 |
| Liberia | 59 | 9300 |
| Syria | 68 | 1200 |
| UK | 76 | 300 |
| Venezuela | 73 | 600 |
| Zambia | 36 | 11 000 |

**FIGURE 1.11**

**3** *Look at 1G.*
Draw a scattergraph to show the information in Figure 1.11.

**4** *Look at 1H.*
(a) Trace the map of Australia shown in Figure 1.12.
(b) Draw a dot distribution map to show the population distribution in Australia, using the information in Figure 1.13 below.

**5** *Look at Figure 1.10.*
Which table of information (A, B or C) could be shown by (a) a line graph, (b) a population pyramid?
Give a reason for each answer.

| AUSTRALIAN STATE | POPULATION (2001) |
|---|---|
| Capital Territory | 300 000 |
| New South Wales | 6 400 000 |
| Northern Territory | 200 000 |
| Queensland | 3 700 000 |
| South Australia | 1 500 000 |
| Tasmania | 500 000 |
| Victoria | 4 600 000 |
| Western Australia | 1 900 000 |

**FIGURE 1.13**

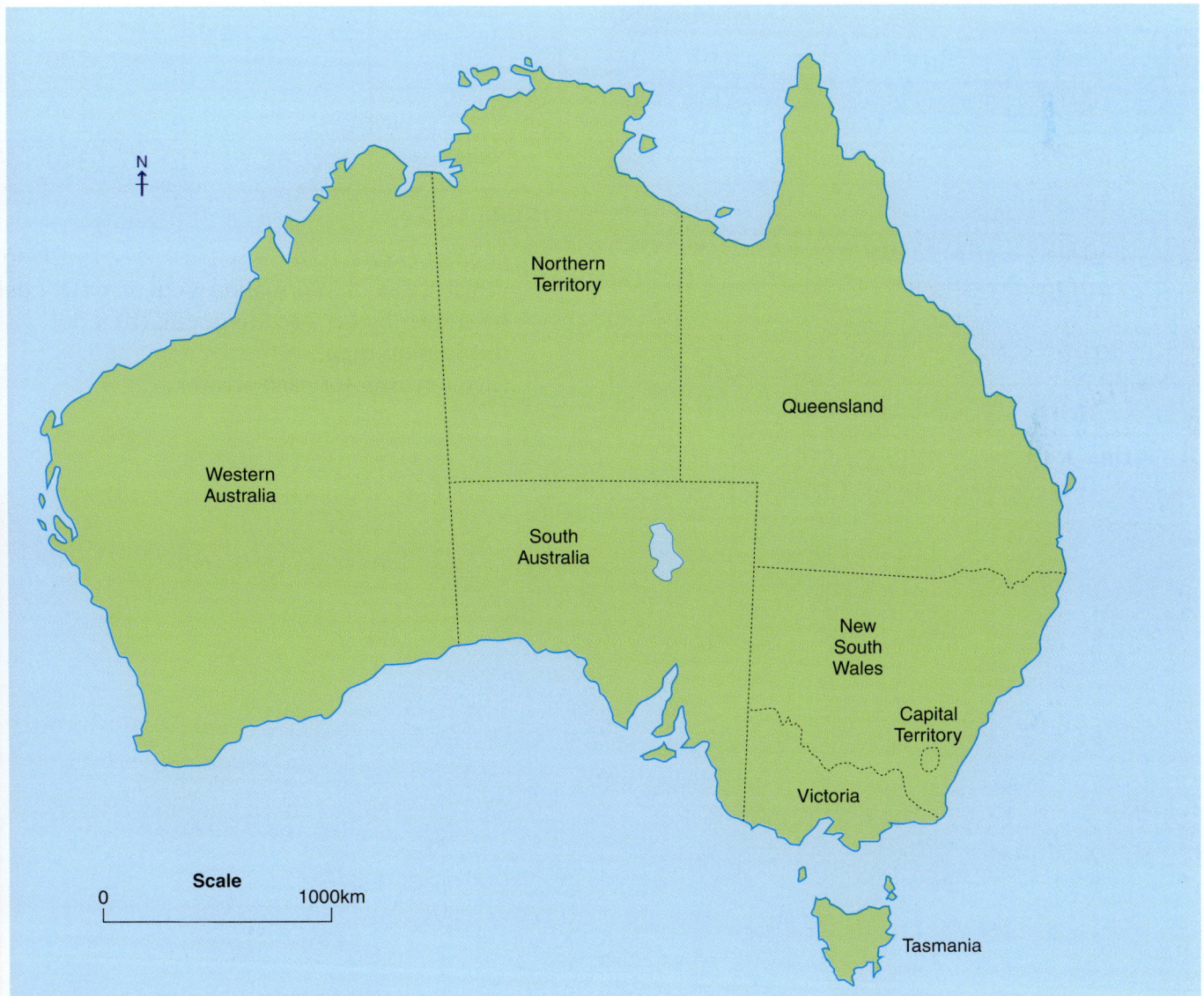

**FIGURE 1.12** *Outline map of Australia*

1. **Look at 1B.**
   What techniques would you use to find out the standard of living of people in two different districts of Scotland? Justify your choice of techniques.

2. **Look at 1C.**
   Describe how you would carry out a questionnaire in a village to find out the population structure (the number of males and females of different ages).

3. **Look at 1F.**
   Draw a population pyramid to show the information in Figure 1.14 below.

| AGE GROUPS | KENYA | |
|---|---|---|
| | MALES (%) | FEMALES (%) |
| 0–9 | 18 | 18 |
| 10–19 | 11 | 11 |
| 20–29 | 7 | 8 |
| 30–39 | 4 | 5 |
| 40–49 | 3 | 3 |
| 50–59 | 2 | 3 |
| 60–69 | 2 | 2 |
| 70+ | 1 | 2 |

**FIGURE 1.14**

4. **Look at 1G.**
   Draw a scattergraph to show the information in Figure 1.15.

| COUNTRY | URBAN POPULATION (%) | GNP PER PERSON (£) |
|---|---|---|
| Jordan | 72 | 940 |
| Malawi | 13 | 110 |
| Malaysia | 52 | 2430 |
| Mongolia | 60 | 190 |
| Netherlands | 89 | 15 000 |
| New Zealand | 86 | 9000 |
| Tunisia | 57 | 1140 |
| UK | 90 | 11 700 |
| Vietnam | 20 | 150 |
| Zimbabwe | 32 | 340 |

**FIGURE 1.15**

5. **Look at Figure 1.10**
   Which table of information (A, B, C or D) could be shown by (a) a scattergraph, (b) a dot distribution map?
   Give a reason for each answer.

# Counting the People

## Core text

### 2A Population

The number of people living in a country is called its **population**. As Figure 2.1 shows, Scotland has a population of five million. The United Kingdom has over ten times more people than Scotland. Europe has ten times more people than the United Kingdom and, in the whole world, there are ten times more people than in Europe.

### 2B A census

We know how many people there are in the world because every country takes a **census**. A census is a count of the number of people living in a country. In Britain we take a census every 10 years and the last one was in 2001. Every household was given a **census form** to fill in on exactly the same date – Sunday 29 April. The forms were then collected and the figures added up. The census form has questions about each person in every household, in order to find out:

(a) **basic facts** – their age, sex, nationality
(b) **extra facts** – other useful details such as occupation, type of house, language spoken

### 2C Why censuses are taken

The governments of every country find it helpful to take a census, because:

1 It tells them how many people live in different parts of the country, so they know how much money should be spent in each region.
2 It tells them how fast the population is growing, so that they can try to slow it down or speed it up, if necessary.
3 It tells them how many schools will be needed, how much money they will need for pensions and how many people will be taxpayers.

All this information is needed for a government to be able to **plan ahead**, for example how many new hospitals, houses and roads should be built and where in the country they should be built.

| REGION | POPULATION |
|---|---|
| Scotland | 5 million |
| United Kingdom | 58 million |
| Europe | 600 million |
| The world | 6000 million |

**FIGURE 2.1**

### 2D Census problems

Although there are good reasons for taking censuses they are rarely accurate, especially in poorer countries. These are some of the reasons why:

1 They are expensive to carry out.
2 It is difficult to reach some villages.
3 Some people cannot read or write and so cannot fill in the forms.
4 It is difficult to count people if there is a war.
5 Some people are nomads (they have no fixed home) and so are difficult to track down.
6 In some countries, many different languages are spoken.
7 Some people do not want to tell the truth.

People must fill in census forms or they may be fined. Even so, the census results in the United Kingdom are not totally reliable.

## 2E Population density and population distribution

We can use censuses to find out the countries with the highest populations in the world. These are shown in Figure 2.2. However, although these countries have a lot of people, they are not necessarily the most crowded countries. To work out how crowded a country is, you must also take into account its area. For example, China has a huge population, but it also covers a huge area. Overall, it is not as crowded as the United Kingdom. **Population density** describes how crowded a country is. To work out population density, you use the formula:

$$\text{population density (people per sq km)} = \frac{\text{population of country}}{\text{area of country (sq km)}}$$

Figure 2.3 shows the ten countries with the highest population density in the world.

Figure 2.4 shows where the crowded areas of the world are found and where the empty areas are found. So it shows the **population distribution**, the spread of people over an area. Most people in the world live in the northern hemisphere, south of the Arctic Circle and in coastal areas.

| THE 10 MOST POPULOUS COUNTRIES | |
|---|---|
| 1  China | 6  Russia |
| 2  India | 7  Pakistan |
| 3  USA | 8  Japan |
| 4  Indonesia | 9  Bangladesh |
| 5  Brazil | 10  Nigeria |

**FIGURE 2.2**

| THE 10 MOST CROWDED COUNTRIES | |
|---|---|
| 1  Bangladesh | 6  Belgium |
| 2  Taiwan | 7  Japan |
| 3  Netherlands | 8  India |
| 4  South Korea | 9  Lebanon |
| 5  Puerto Rico | 10  Sir Lanka |

**FIGURE 2.3**

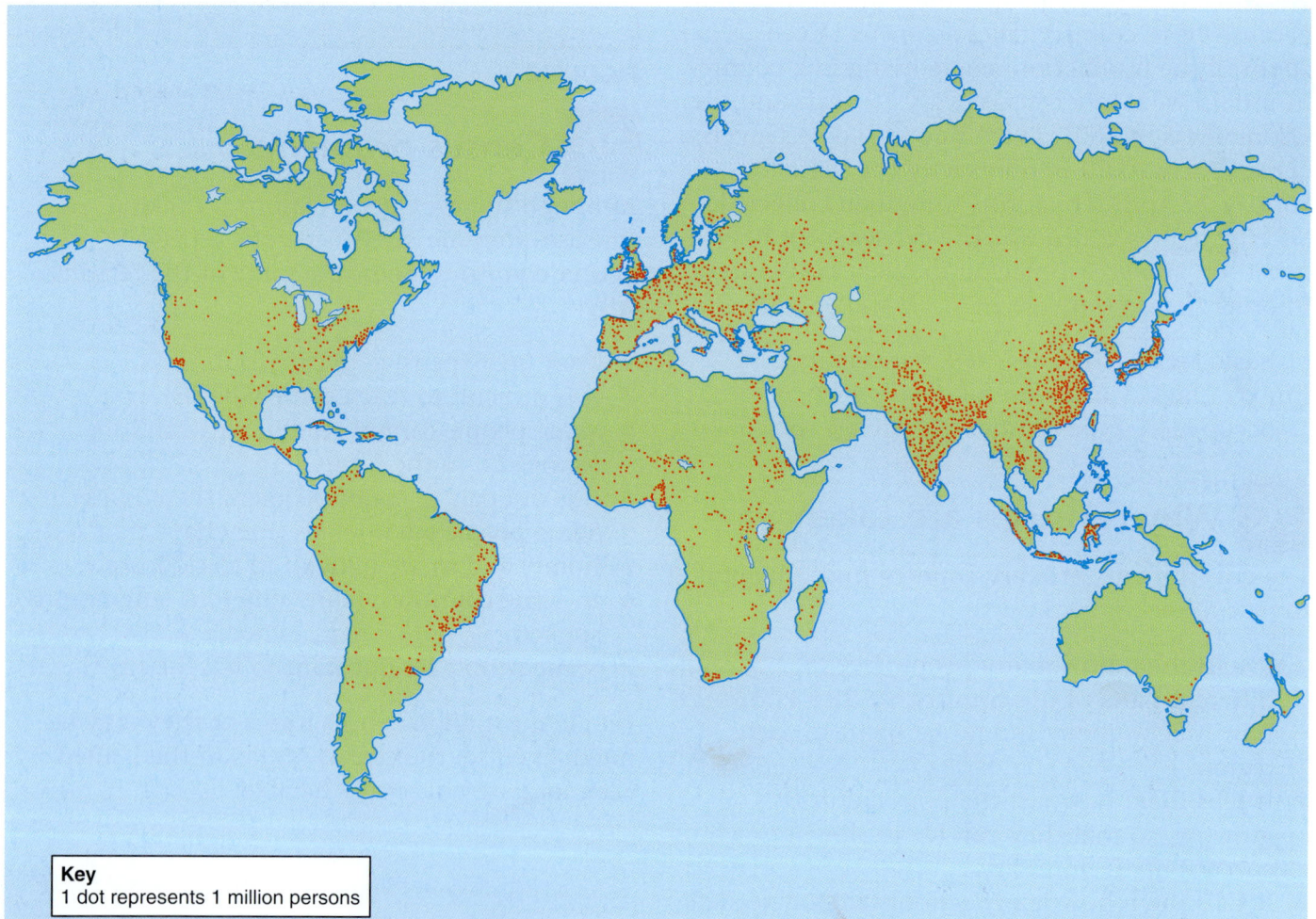

Key
1 dot represents 1 million persons

**FIGURE 2.4** *Distribution of the world's population*

## 2F  Population structure

A census will also tell us the make up of a population, for example how many males and females there are, of different ages. This is called the **population structure** and is usually shown by a population pyramid. Figure 2.5 is a population pyramid showing the world's population structure. The largest age group is the 0–4 year olds and the smallest is the 70–74 year olds. The number of males and females in each age group is very similar except for the over 70s, where there are far more females.

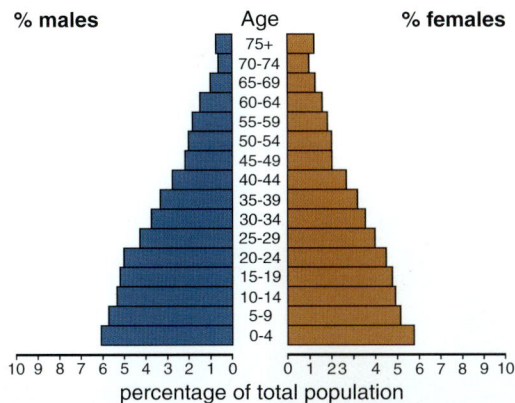

**FIGURE 2.5** *Population structure of the world*

## CORE QUESTIONS

1. *Look at 2A.*
   What is the population of (a) Scotland, (b) the world?

2. *Look at 2B.*
   What is a census?

3. How often does Britain take a census?

4. *Look at 2C.*
   Give two reasons why censuses are taken.

5. *Look at 2D.*
   Why do poor countries find it difficult to take censuses?

6. Why are nomads difficult to count in a census?

7. Why is it much harder to carry out a census if the people cannot read or write?

8. *Look at Figure 2.2.*
   Which two countries have the most people in the world?

9. *Look at 2E.*
   What does it mean if a country has a high population density?

10. What is meant by the population distribution in an area?

## FOUNDATION QUESTIONS

### Case Study of Nigeria

1. *Look at Figure 2.11.*
   Where do (a) most people and (b) fewest people in Nigeria live?

2. *Look at Figure 2.12.*
   Which age group in Nigeria has (a) the most males and females, and (b) the fewest males and females?

3. *Look at Figure 2.9.*
   How did the population of Nigeria change between 1962 and 1991?

4. *Look at Figure 2.10.*
   How do we know that the 1962 and 1963 census figures were wrong?

5. In what way have the leaders of the main tribal groups made censuses inaccurate?

6. Which of the two facts below make it more difficult to take an accurate census? Give reasons for your answer.

   **Fact 1:** Nigeria has some land covered in rainforest.
   **Fact 2:** Many different languges are spoken in Nigeria.

7. How much did the 1991 census cost?

8. *Look at Figure 2.13.*
   In what ways does the Nigerian government use census information?

## Case Study of Nigeria

**1** *Look at Figure 2.11.*
Where are (a) the most crowded, and (b) the least crowded parts of Nigeria?

**2** *Look at Figure 2.12.*
Compare the number of people in different age groups in Nigeria.

**3** *Look at Figure 2.9.*
Describe the changes in Nigeria's population between 1962 and 1991.

**4** *Look at Figure 2.10.*
How can we tell from the results that the census figures were inaccurate?

**5** Why did the leaders of the main tribal groups try to make their populations higher than they actually were?

**6** Why was the 1991 census so expensive?

**7** *Look at Figure 2.13.*
Suggest why Nigeria continues to take censuses.

**FIGURE 2.6** *Dense rainforest in southern Nigeria*

# RESOURCES

## Case Study of Nigeria

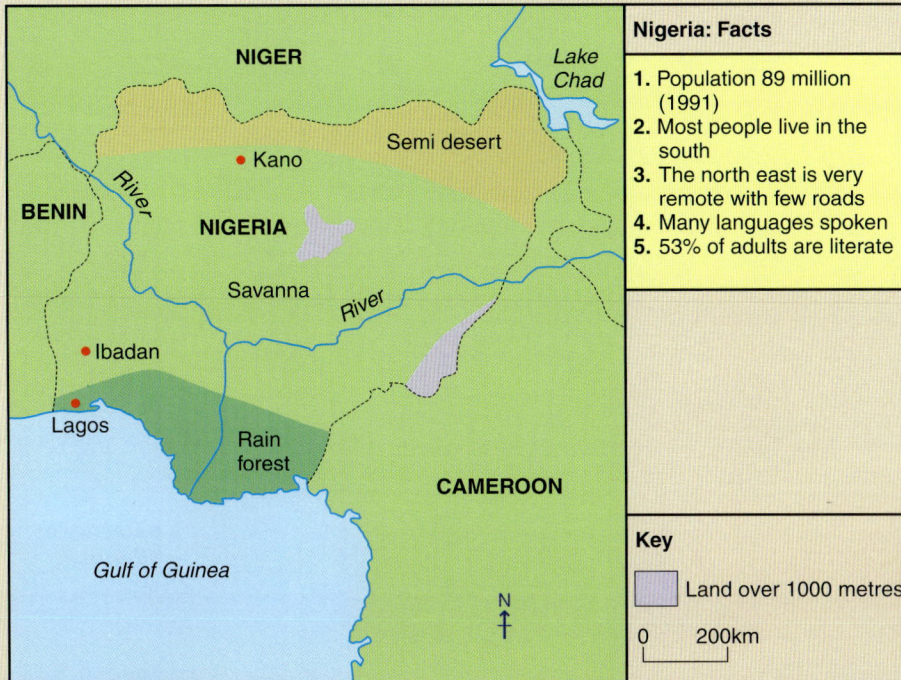

**Nigeria: Facts**

1. Population 89 million (1991)
2. Most people live in the south
3. The north east is very remote with few roads
4. Many languages spoken
5. 53% of adults are literate

**Key**

◻ Land over 1000 metres

0    200km

**FIGURE 2.7** | *Nigeria*

### INTRODUCTION

Nigeria is a large country in West Africa. It is an old colony of Britain, which became independent in 1960. Since then, it has taken four censuses, but the results have proved to be inaccurate. The most recent one, in 1991, is thought to be the most reliable. It shows Nigeria to have a population of 89 million. This means it has the largest population of any country on the continent of Africa.

**FIGURE 2.8**

### CENSUS RESULTS

| 1991 | 89 million |
|------|------------|
| 1973 | 80 million |
| 1963 | 55 million |
| 1962 | 45 million |

**FIGURE 2.9**

### CENSUS PROBLEMS

Figure 2.9 shows the most recent census results in Nigeria and they look very suspicious. It is not possible for a country's population to increase naturally by 10 million (22 per cent) in one year, as Nigeria's did between 1962 and 1963. It is also very unlikely that it can increase by 25 million (45 per cent) in ten years, and then only by another 9 million in the next 18 years.

One reason why censuses have been inaccurate in the past is because the four main tribal groups in different parts of Nigeria tried to make their population totals higher than they actually were. They did this in order to get more money for their region and more votes in the Nigerian parliament. Other census problems are shown in Figure 2.7.

To try and make the 1991 census accurate, Nigeria employed 500 000 officials. They also used many computers to collect and sort the information and to detect cheating. Several countries helped to pay the £75 million that the census cost.

**FIGURE 2.10**

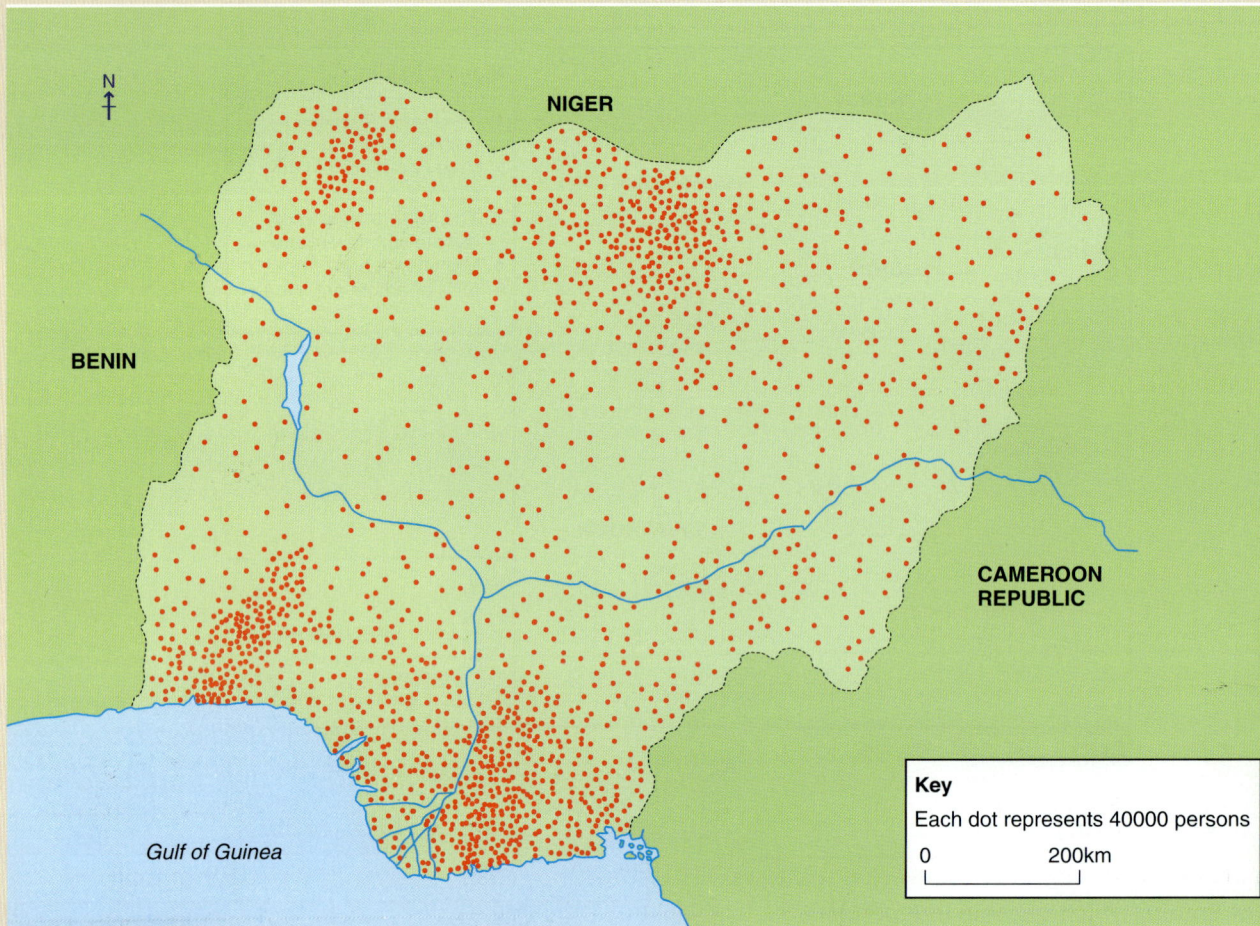

**FIGURE 2.11** *Distribution of population in Nigeria*

NIGER

BENIN

CAMEROON REPUBLIC

Gulf of Guinea

Key

Each dot represents 40000 persons

0     200km

## POPULATION CHARACTERISTICS

Figures 2.11 and 2.12 show the population distribution and the population structure in Nigeria. These have been worked out from the 1991 census. With all this census information, Nigeria was able to plan its future better. It could work out how many new schools and hospitals to build and where to build new roads and new houses.

**FIGURE 2.13**

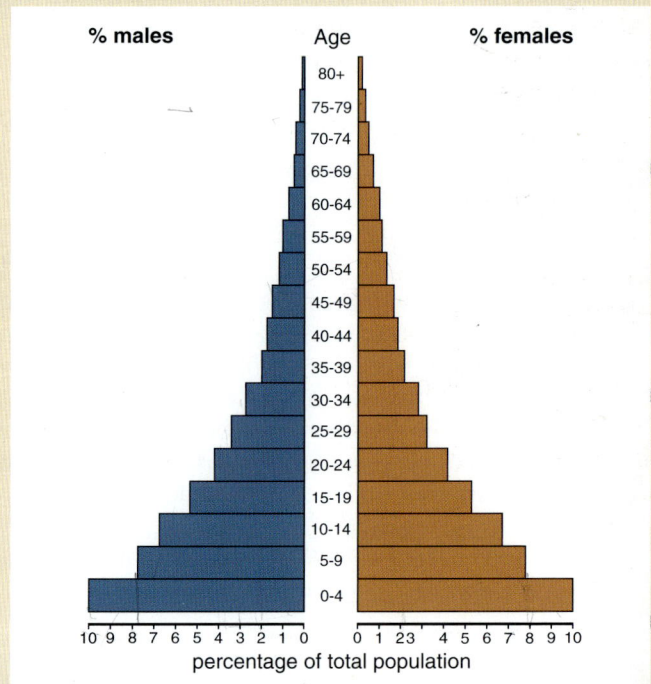

% males          Age          % females

80+
75-79
70-74
65-69
60-64
55-59
50-54
45-49
40-44
35-39
30-34
25-29
20-24
15-19
10-14
5-9
0-4

10 9 8 7 6 5 4 3 2 1 0    0 1 2 3 4 5 6 7 8 9 10

percentage of total population

**FIGURE 2.12** *Nigeria's population structure*

## 2G  More Census problems

The United Nations would like every country in the world to take a census every ten years, but some countries still have not taken a full census. Not only is the information useful to governments, it is increasingly used by companies wanting to know more about the make-up of the population in an area before deciding to locate their factories and offices there.

However, censuses are often unreliable. It also takes a long time to process all the information, by which time some of it is already out-of-date. They are also expensive to carry out. The 2001 census in Britain cost £255 million.

Because of all the difficulties some countries, such as Sweden, have now stopped taking censuses. They believe there are better ways of counting people.

## 2H  Vital registrations

A census is not the only way of counting people. Records can be kept of vital events such as births, deaths, marriages, adoptions and divorces.

In Britain, these **vital registrations** are compulsory. Without them people would find it very difficult to get passports, health certificates, life insurance and other things. This makes them a reliable way of counting people. In poorer countries they are unreliable because they are not compulsory, many people are illiterate and live a long way from the registration offices. Vital registrations provide a continuous record of population changes and are quite cheap to administer. A census, on the other hand, records many details of the population at one point in time and it is expensive to undertake.

## >> EXTENSION QUESTIONS

*Look at the Extension Text.*

1. Explain why some countries have stopped taking censuses.

2. What are 'vital registrations'?

3. Explain why vital registrations in poorer countries are unreliable.
   '*A census is like a photograph. Vital registrations are like a film*'.

4. Explain what the statement above means.

## CREDIT QUESTIONS

### Case Study of Nigeria

1. *Look at Figure 2.11.*
   Describe the distribution of population in Nigeria.

2. *Look at Figure 2.12.*
   Describe the population structure in Nigeria.

3. *Look at Figure 2.10.*
   Explain why some people in Nigeria deliberately try to make censuses inaccurate.

4. *Look at Figure 2.7.*
   Suggest other problems Nigeria faces in taking accurate censuses.

5. *Look at Figures 2.10 and 2.13.*
   Describe the different points of view political leaders in Nigeria might have towards taking another census in the near future.

# 3 World Population Distribution

## Core text

### 3A Crowded areas and empty areas

The world is not overcrowded. It is not even crowded. There are, on average, only 40 people to every square kilometre of land. Some areas, of course, are much more crowded than this (see Figure 3.1), but there are also a lot of empty areas (see Figure 3.2). The world's population is scattered very unevenly. One small area of north east USA (see Figure 3.1) contains more people than the huge countries of Canada and Australia together. There are many reasons for this uneven distribution and they are studied separately in this unit.

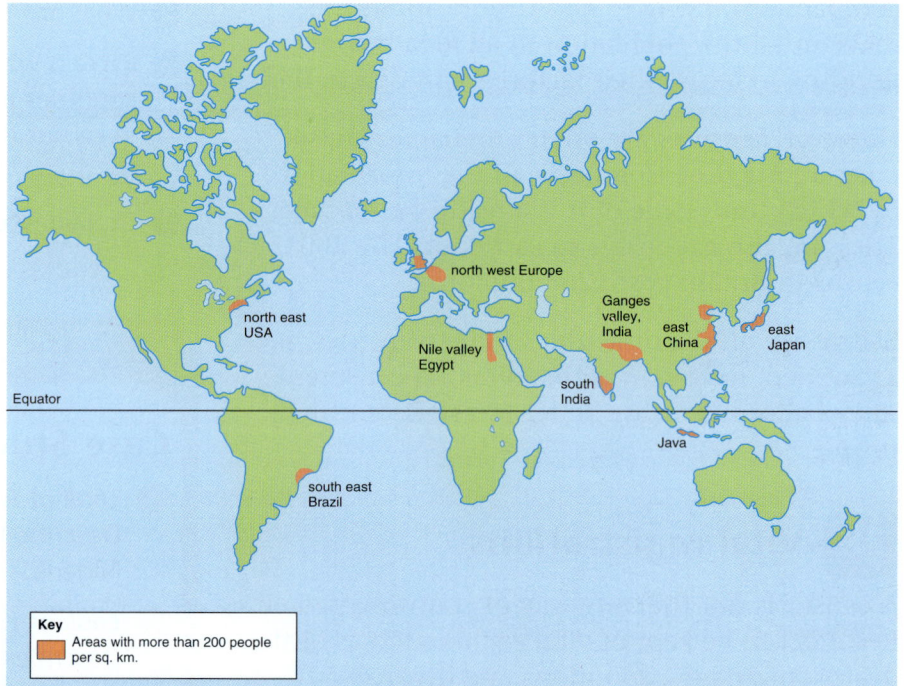

**north west Europe**
**north east USA**
**Ganges valley, India**
**east China**
**east Japan**
**Nile valley Egypt**
**south India**
**Java**
**south east Brazil**
**Equator**

**Key**
Areas with more than 200 people per sq. km.

**FIGURE 3.1** *The most crowded areas of the world*

### 3B The importance of climate

**Polar regions** have few people because:

- it is difficult and expensive to survive in such cold conditions

- it is impossible to grow crops, so food is expensive

- building is difficult because of the permafrost underneath

- they are very remote, due to the poor weather and lack of roads and railways.

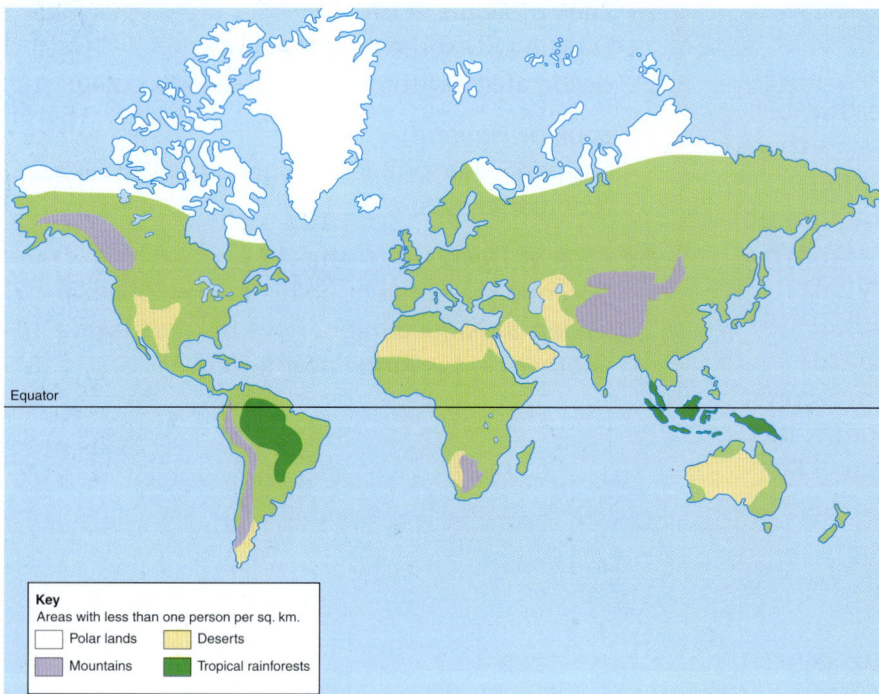

**Equator**

**Key**
Areas with less than one person per sq. km.
- Polar lands
- Deserts
- Mountains
- Tropical rainforests

**FIGURE 3.2** *The most empty areas of the world*

**Very dry regions (hot deserts)** have few people because:

- living in extreme heat and drought is unpleasant for most people

- it is impossible to grow crops without irrigation, so food is expensive

- they are remote and unlikely to attract industries, so there are few job opportunities.

**Regions with a moderate climate and reliable water supply** are more crowded because:

- it is a comfortable climate in which to live

- no extra costs are needed in order to cope with the climate

- farmers can have steady harvests from year to year because the climate is reliable.

## 3C The importance of relief

**Steep, mountainous regions** have a low population density because:

- they are very cold (except in the Tropics) and so living and farming conditions are harsh

- it is too steep for roads and railways, so the areas are remote

- it is too steep on which to build houses and factories and to use farm machinery, so job opportunities are few.

**Flat and gently sloping areas** have a high population density because:

- it is easy to build houses and factories

- farming is more productive because the soil is deeper and machinery can be used

- roads and railways can be built, which encourages industry.

## 3D The importance of soils and vegetation

**Tropical rainforests** have a low population density because:

- the hot, humid climate is uncomfortable in which to live

- dense forest is difficult to clear in order to build houses, roads etc.

- soils are very poor, once the trees have been cut down.

**Areas with poor soils** have a low population density because:

- crops grow so badly that the farms have to be very large for the farmer to make a profit.

**Areas with fertile soils** have a higher population density because:

- crops grow very well and so farms are smaller.

## 3E The importance of economic activities

**Farming areas** usually have a low population density because:

- each family needs a large area to make a living (unless the soil is extremely fertile).

**Industrial areas** have a high population density because:

- factories and offices only take up a small area, yet provide many jobs.

**FIGURE 3.3** *The empty areas of the Himalayas*

**FIGURE 3.4** *The crowded areas of New York*

## 3F The importance of development

**Areas with a low level of development** have few people because:

- farming methods will be poor and there is little farm equipment, so farms need to be quite large

- there are few factories and offices, as there are not enough skilled workers, managers or capital available.

**Areas with a high level of development** are more crowded because:

- most people work in offices and factories which need large populations nearby.

### CORE QUESTIONS

1. *Look at Figure 3.1.*
   Give three examples of crowded areas in the world.

2. *Look at Figure 3.2.*
   Give three examples of empty areas in the world.

3. *Look at 3B.*
   Give two reasons why (a) polar regions and (b) hot deserts have few people.

4. *Look at 3C.*
   Why are mountainous areas often remote?

5. Give two reasons why flat areas have a high population density.

6. *Look at 3D.*
   Give two reasons why tropical rainforests have few people.

7. In what way does the soil affect population density?

8. *Look at 3E.*
   Why are industrial areas densely populated?

9. *Look at 3F.*
   Why do areas with a low level of development have fewer people?

## FOUNDATION QUESTIONS

**1** *Look at Figure 3.8.*
Which is (a) the most densely populated, and (b) the least densely populated area of Norway?

**2** *Look at Figure 3.9.*
Why is it difficult to farm in north Norway?

**3** In what ways has the government helped to increase the number of people living in north Norway?

**4** *Look at Figure 3.10.*
Why is Trondelag more crowded than north Norway?

**5** *Look at Figure 3.11.*
Suggest why it is difficult to build roads and railways in west Norway.

**6** Do you think west Norway is a popular place to live? Give reasons for your answer.

**7** *Look at Figure 3.12.*
In what ways does the high, steep land of south Norway attract people to live there?

**8** *Look at Figure 3.13.*
What, do you think, is the most important reason why Oslofjord is the most crowded area of Norway? Give reasons for your answer.

## GENERAL QUESTIONS

**1** *Look at Figure 3.8.*
Compare the population density in the different regions of Norway.

**2** *Look at Figure 3.9.*
What, do you think, is the most important reason why north Norway is sparsely populated? Give reasons for your answer.

**3** In what ways has the government improved conditions for the people in north Norway?

**4** *Look at Figure 3.11.*
Describe in detail, one advantage and one disadvantage of living in west Norway.

**5** *Look at Figure 3.12.*
Do you think the climate and relief of south Norway has made it a popular place to live? Give reasons for your answer.

**6** *Look at Figure 3.13.*
What is the most important reason why Oslofjord is densely populated? Give reasons for your answer.

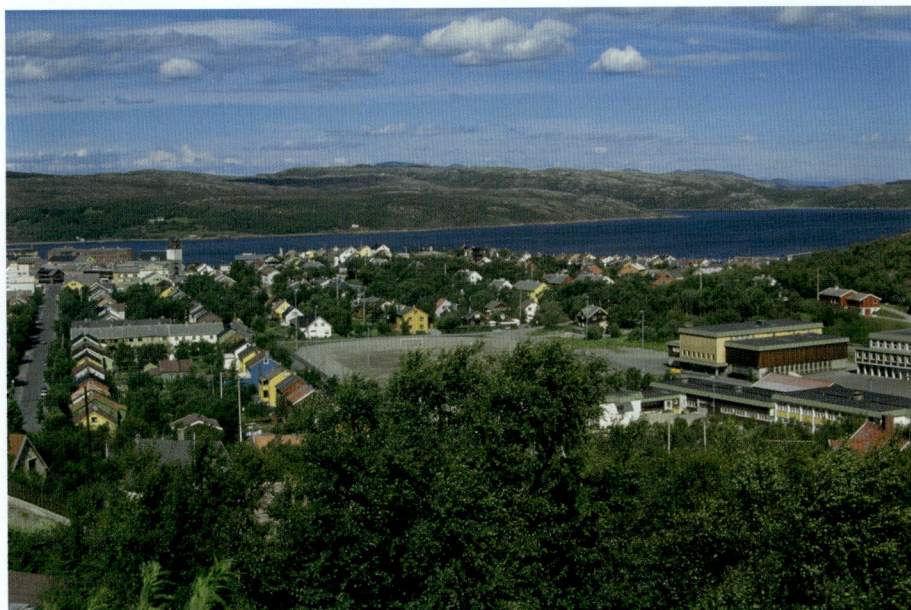

**FIGURE 3.5** *Remote northern Norway, 2000 km from Oslo*

# RESOURCES
## Case Study of Norway

### INTRODUCTION

Norway is one and a half times bigger than the UK, but the UK has nearly 15 times as many people. Norway is the least densely populated country in Europe. Its small population is not evenly scattered over the whole country, as you can see in Figure 3.7. There are large areas where few people live, but there are also small pockets of high population density.

**FIGURE 3.6**

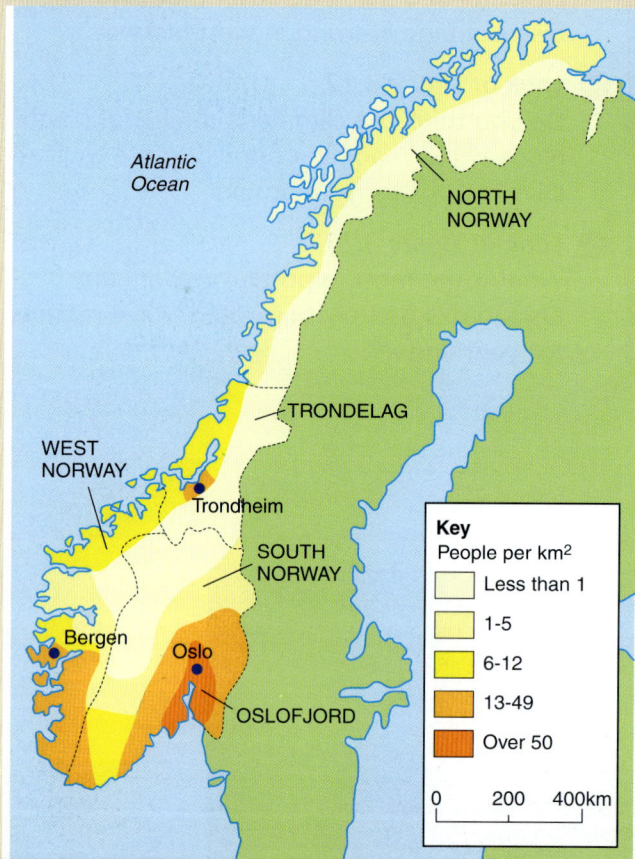

**FIGURE 3.7** *Density of population in Norway*

### NORTH NORWAY

North Norway is the emptiest region. Its main problem is its remoteness. The most northerly settlement, Kirkenes, is nearly 2500km from the capital, Oslo. Being so far away from the rest of Norway makes it less likely to attract industries and people to the area. Another problem is that most of the area lies within the Arctic Circle. Although winters are not very cold (just below 0°C), the growing season is short and only 1 per cent of the area is cultivated.

The small population live along the coast, where there are some jobs in the small ports, in fishing and in services. The government has given subsidies to attract other industries to the area, such as steelworks, aluminium works and iron ore mining. It has built small airfields near the largest settlements, so that the area is no longer quite as remote.

**FIGURE 3.9**

### TRONDELAG

This region almost reaches the Arctic Circle, but it has a much higher population density and contains the third largest town in Norway, Trondheim. It has a higher population density chiefly because it has fertile soils which, along with the long hours of summer daylight, and flatter, lower land, mean that a variety of crops can be grown. Barley is the most important crop but dairying also takes place.

**FIGURE 3.10**

| REGION | AREA (THOUSAND KM²) | POPULATION (THOUSANDS) | POPULATION DENSITY (PEOPLE/KM²) |
|---|---|---|---|
| north Norway | 108 | 500 | 5 |
| Trondelag | 41 | 400 | 10 |
| west Norway | 61 | 1100 | 18 |
| south Norway | 94 | 800 | 9 |
| Oslofjord | 11 | 1500 | 136 |

**FIGURE 3.8**

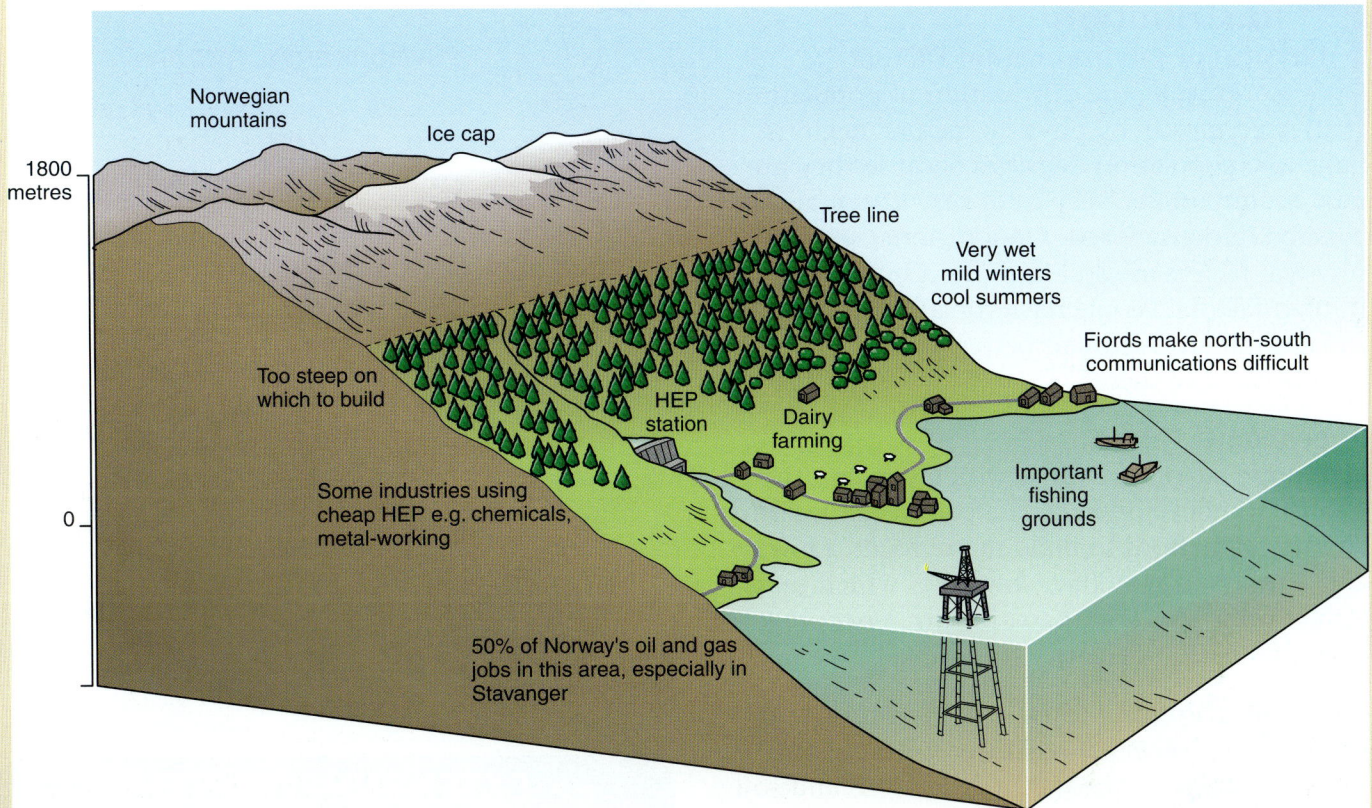

Labels on diagram:
- Norwegian mountains
- Ice cap
- 1800 metres
- Tree line
- Very wet mild winters cool summers
- Fiords make north-south communications difficult
- Too steep on which to build
- HEP station
- Dairy farming
- Important fishing grounds
- Some industries using cheap HEP e.g. chemicals, metal-working
- 0
- 50% of Norway's oil and gas jobs in this area, especially in Stavanger

**FIGURE 3.11** *Block diagram of a typical west Norway landscape*

## SOUTH NORWAY

This area contains the highest mountains, which reach over 2400 metres and have ice caps and glaciers on them. It is popular with tourists in winter for skiing (for example at Lillehammer) and in summer because of the warm temperatures and numerous beaches. South Norway is the most important region in the country for forestry and for the production of hydroelectricity. The steep slopes make farming and building difficult but they are excellent for HEP and the cheap power has attracted metal-working industries. The trees that grow on the steep slopes are used in the timber industry here. The only large town, Kristiansand, has iron and metal industries and shipbuilding.

**FIGURE 3.12**

## OSLOFJORD

This is by far the most densely populated region of Norway. It contains the capital and biggest city, Oslo, which is the country's chief port and industrial centre. In addition, the area has the most productive farmland, containing the most fertile soils. It is a lowland area with the warmest summers in Norway, although the growing season is short and snow lies on the ground for over 100 days a year.

**FIGURE 3.13**

## 3G Factors in population distribution

### 1 Physical or Environmental Factors

On a world scale, the distribution of population is affected mostly by physical factors. Regions have low population densities because they are very mountainous, very cold, very dry or are covered in dense forest. On a regional scale, physical factors are still important but so are human factors. People have the ability to make the natural environment better, so that more people can live there.

### 2 Economic Factors

Economic factors are those which concern the wealth and development of a country. A wealthy country can afford to build many roads, railways and airports. It may have low taxes which may attract industries and these, in turn, encourage a high population density.

A developed country can use its technology to overcome problems of relief (for example by tunnels, bridges), climate (by irrigation) and soil (by fertilisers). Technology makes it possible for a lot of people to live in an area which otherwise would be sparsely populated.

If the people in a country are highly skilled and educated, they will attract industries and offices, which mean that many people can live in a small area.

### 3 Political Factors

Political factors are those which concern governments and the decisions they make. If the government of a country decides to invest a lot of money and technology in one region (for example building roads, irrigation schemes, New Towns), more people will go to live in that area and the population density will rise.

## 3H Summary

**ENVIRONMENTAL FACTORS**

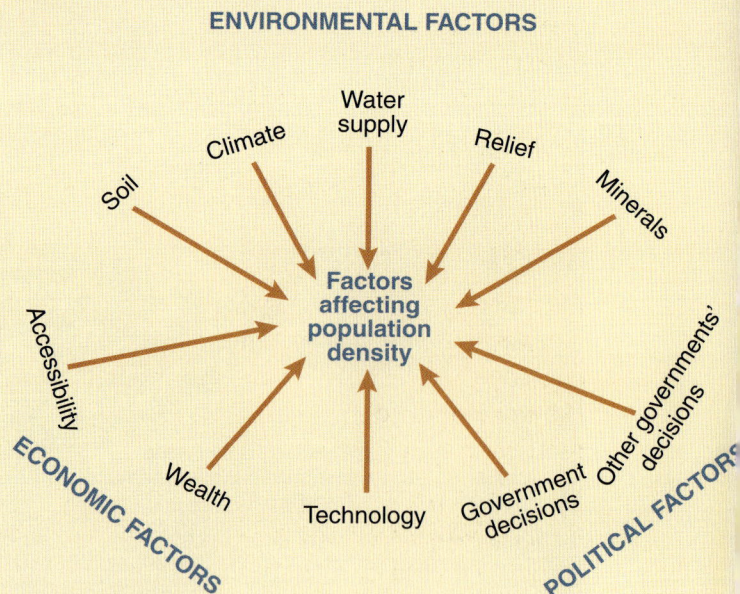

**FIGURE 3.14** *Factors affecting population distribution*

## >> EXTENSION QUESTIONS

*Look at the Extension Text.*

1. What are the main physical factors affecting population distribution?

2. Give examples of the ways in which people can overcome problems of the physical environment.

3. In what ways can a rich country support a greater population density compared to a poor country?

4. In what ways do governments affect population density?

## CREDIT QUESTIONS

1. *Look at Figures 3.7 and 3.8.*
   Describe the relationship between population density in Norway and latitude.

2. *Look at Figure 3.9.*
   In what ways have political factors affected population density in north Norway?

3. *Look at Figure 3.11.*
   Describe the advantages and disadvantages of the physical environment to the people of west Norway.

4. *Look at Figure 3.12.*
   Norway is a rich and highly developed country. Do you think that this has affected the population density in south Norway? Give reasons for your answer.

5. *Look at Figure 3.13.*
   Suggest why Oslofjord is the most densely populated area of Norway.

# 4 Standard of Living

## Core text

### 4A Standard of living

Censuses tell us how many people live in a country and they can be used to work out the population density. Censuses can also tell us how well off the people are. This is called their **standard of living**. People who are not well off have a low standard of living. People who are well off have a high standard of living.

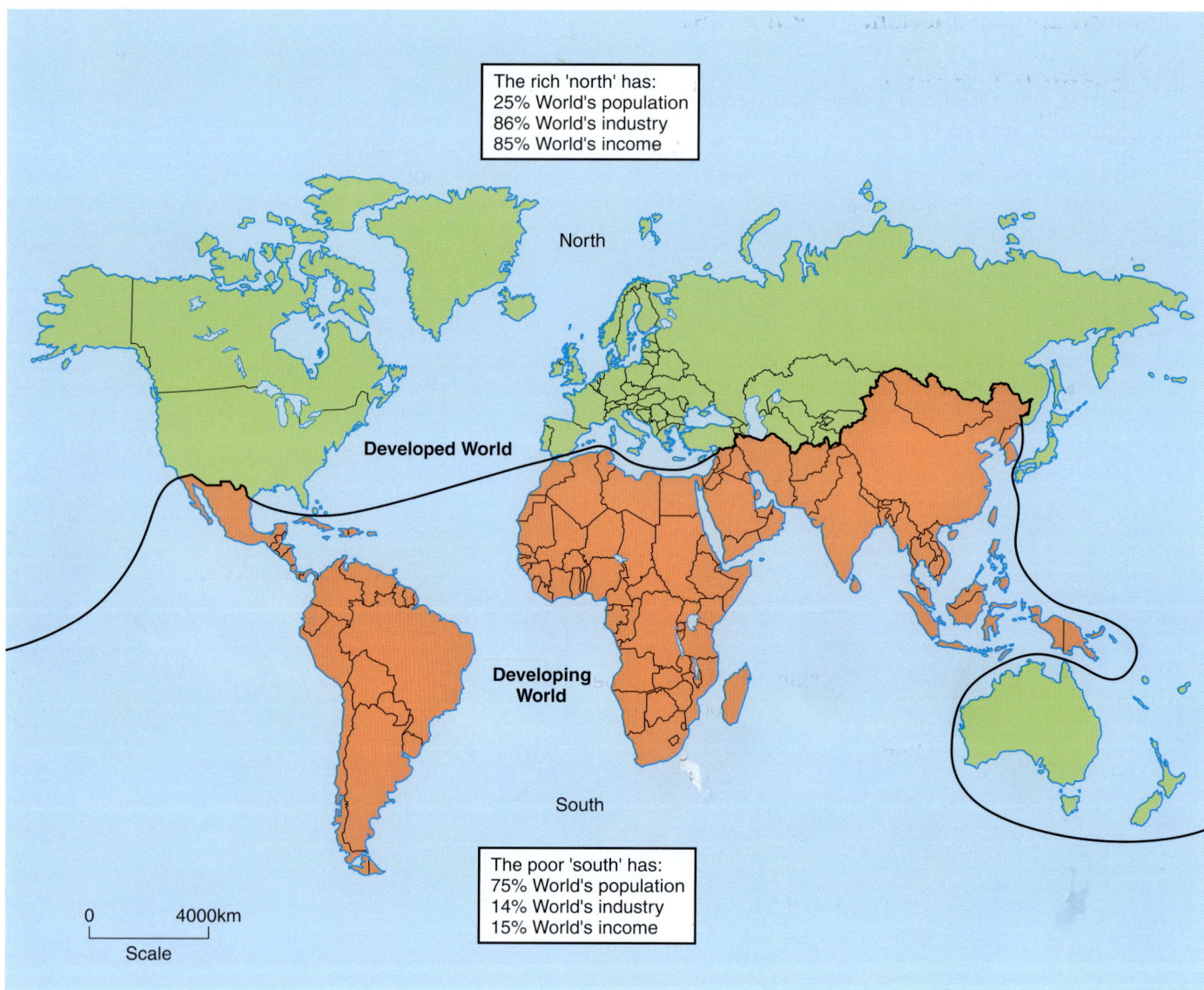

The rich 'north' has:
25% World's population
86% World's industry
85% World's income

North

Developed World

Developing World

South

The poor 'south' has:
75% World's population
14% World's industry
15% World's income

0    4000km
Scale

**FIGURE 4.1** *The developed and developing worlds*

## 4B Developed and developing countries

Countries with a high standard of living are called **developed countries**. They are sometimes called '**the North**' because they are found mainly in the northern hemisphere. Countries with a low standard of living are called **developing countries**. They are sometimes called '**the South**' or the **third world**. Figure 4.1 shows the location of the developed and developing countries in the world.

## 4C Measuring standard of living

The most common way of measuring the standard of living is by working out the wealth of a country. This is done by adding up the value of the goods and services produced in a country in one year. This is called the **gross national product (GNP)**. The GNP is then divided by the number of people in that country, to find out how much wealth the 'average' person receives (**GNP/person or GNP/capita**). This is one **indicator of standard of living**, but there are others. For people to have a high standard of living, they must have more than wealth. They also need to be healthy, educated, well fed and have access to modern technology. These are also indicators of standard of living and are shown in Figure 4.2.

## CORE QUESTIONS

**1** *Look at 4A.*
If a country has a high standard of living, what does this mean?

**2** *Look at 4B.*
What is a developed country?

**3** What is a developing country?

**4** *Look at Figure 4.1.*
Where in the world are the developed countries found?

**5** Where in the world are the developing countries found?

**6** *Look at 4C.*
How is the wealth of a country usually worked out?

**7** How are the following aspects of a country measured:
(a) its education service
(b) its health service
(c) the amount of food available
(d) the amount of industrialisation?

| ASPECT OF STANDARD OF LIVING | INDICATOR OF STANDARD OF LIVING | REASON USED |
|---|---|---|
| Wealth | GNP per capita | Shows how much wealth is available to give people a high standard of living |
| Food | Calories per person per day | Shows how much food is available |
| Health | Number of people per doctor | Show how good the health service is |
| Education | Percentage of people who are literate | Shows how good the education system is |
| Industrialisation | Percentage of people in agriculture | Shows how industrialised the country is |

**FIGURE 4.2** *Indicators of standard of living*

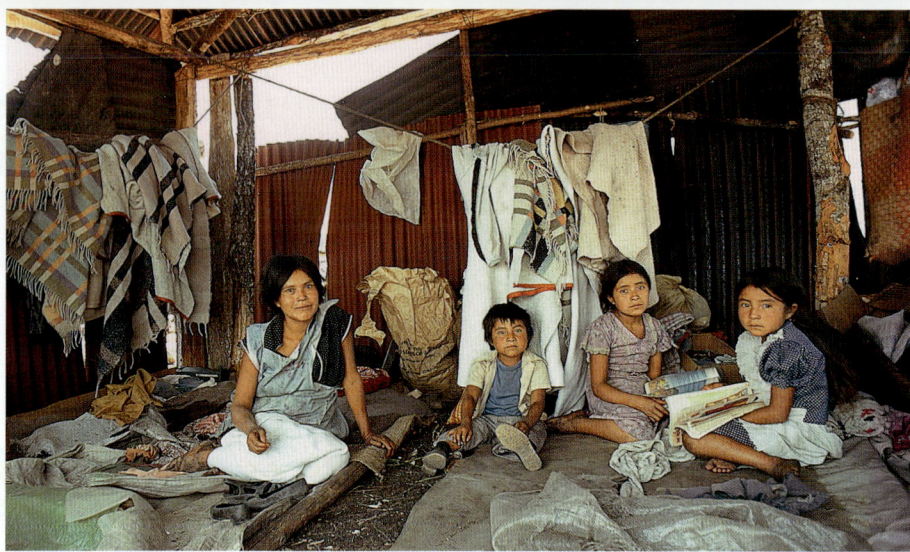

**FIGURE 4.3** *Poor area of Mexico City*

## Case Study of north and central America

1. *Look at Figure 4.6.*
   Of the ten countries in north and central America, which has:
   (a) the highest GNP per person
   (b) the lowest population per doctor
   (c) the lowest percentage of people in agriculture?

2. Which country has the **lowest** standard of living according to:
   (a) GNP per person
   (b) calories per person
   (c) literacy rate?

3. Which country, Guatemala or Honduras, has the higher standard of living? Give reasons for your answer.

4. Which country, Belize or Costa Rica, has the higher standard of living? Give reasons for your answer.

5. Do you think Mexico is a developing country? Give reasons for your answer.

6. *Look at Figure 4.8.*
   Which area of the USA has the highest standard of living?

7. *Look at Figure 4.9.*
   Which areas of the USA have the highest quality of life?

8. *Look at Figure 4.10.*
   Describe the standard of living of people who live in the cities of central America.

## Case Study of north and central America

1. *Look at Figure 4.6.*
   Rank the ten countries of north and central America according to (a) GNP per person, (b) calories per person, and (c) population per doctor (1 = highest standard of living; 10 = lowest standard of living).

2. Which of the ten countries are developing countries? Give reasons for your answer.

3. '*Belize has a higher standard of living than Costa Rica.*'
   Give an argument for and against this point of view.

4. Which has a higher standard of living, Panama or Mexico? Give reasons for your answer.

5. *Look at Figure 4.8.*
   Describe the differences in standards of living across the USA.

6. *Look at Figure 4.9.*
   Describe the differences in quality of life across the USA.

7. *Look at Figure 4.10.*
   Which parts of central America have a higher standard of living, the cities or the countryside? Give reasons for your answer.

# RESOURCES

## Case Study of north and central America

### INTRODUCTION

North and central America stretches from the humid tropics of Panama to the icy wastes of northern Canada. The mainland is made up of ten countries, ranging from the extremely large (such as Canada) to the very small, such as Belize (over 400 times smaller than Canada). This huge region also includes one of the world's richest countries (the USA) and some very poor countries, such as Nicaragua. In Figure 4.6, five indicators have been used to show the differences in the standard of living between these ten countries of north and central America.

**FIGURE 4.4**

### DIFFERENCES WITHIN THE COUNTRIES OF NORTH AMERICA

The USA and Canada are the two richest countries in this region, but this does not mean that everyone is rich. Some areas of the USA and Canada are richer than others and, even in rich cities such as New York, there is great poverty. Figure 4.8 shows the differences in the standard of living within the USA, using average income as the economic indicator. Figure 4.9 shows the differences in quality of life across the USA, using a range of social indicators.

**FIGURE 4.7**

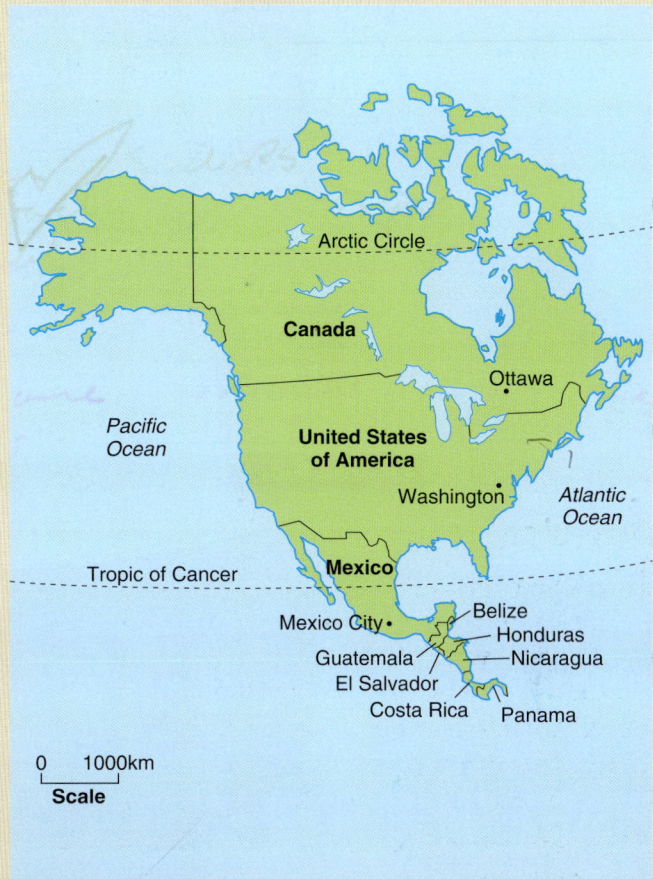

**FIGURE 4.5**  *The countries of north and central America*

| COUNTRY | GNP PER PERSON (DOLLARS) | CALORIES PER PERSON | POPULATION PER DOCTOR | LITERACY RATE | % OF PEOPLE IN AGRICULTURE |
|---|---|---|---|---|---|
| Canada | 19 400 | 3100 | 450 | 99 | 5 |
| USA | 27 000 | 3700 | 420 | 99 | 3 |
| Belize | 2600 | 2700 | 1500 | 96 | 18 |
| Costa Rica | 2600 | 2900 | 1030 | 93 | 25 |
| El Salvador | 1600 | 2700 | 1560 | 70 | 11 |
| Guatemala | 1300 | 2200 | 4000 | 54 | 50 |
| Honduras | 600 | 2300 | 1270 | 71 | 38 |
| Mexico | 3300 | 3200 | 620 | 87 | 23 |
| Nicaragua | 400 | 2300 | 2000 | 65 | 46 |
| Panama | 2800 | 2200 | 560 | 90 | 27 |

**FIGURE 4.6**

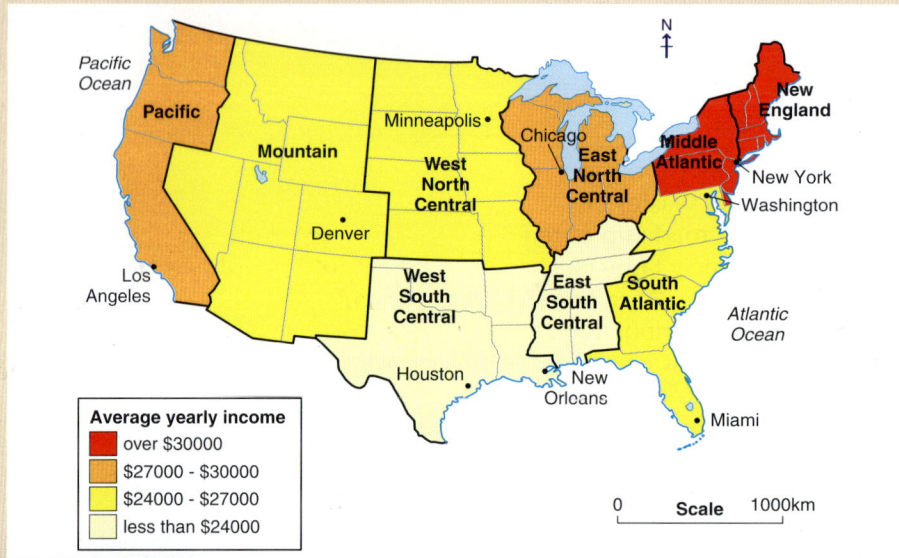

**Average yearly income**
- over $30000
- $27000 - $30000
- $24000 - $27000
- less than $24000

Pacific Ocean · Pacific · Mountain · West North Central · East North Central · Middle Atlantic · New England · Minneapolis · Chicago · New York · Washington · Denver · Los Angeles · West South Central · East South Central · South Atlantic · Atlantic Ocean · Houston · New Orleans · Miami

Scale 0 — 1000km

**FIGURE 4.8** *Differences in the standard of living within the USA*

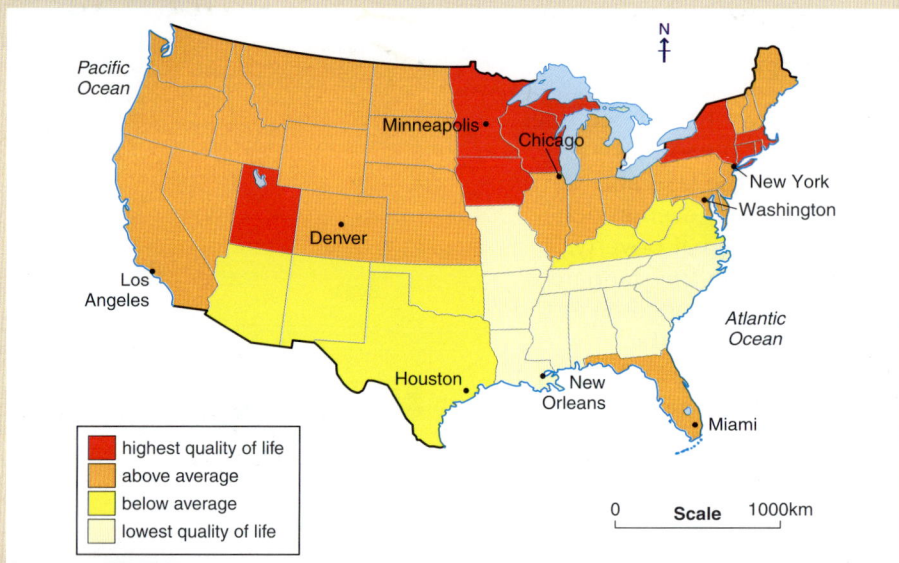

- highest quality of life
- above average
- below average
- lowest quality of life

Pacific Ocean · Minneapolis · Chicago · New York · Washington · Denver · Los Angeles · Houston · New Orleans · Miami · Atlantic Ocean

Scale 0 — 1000km

**FIGURE 4.9** *Differences in quality of life within the USA*

## DIFFERENCES WITHIN THE COUNTRIES OF CENTRAL AMERICA

Differences in the standard of living in the poorer countries of central America are greater than those in the USA and Canada. Most of the wealthy people live in the cities, although there are also many homeless and extremely poor people living in the cities as well. The cities also contain most of the services. Here, you can find universities, hospitals, shops and entertainments. In the countryside, however, there are few well paid jobs and so the standard of living is much lower. A typical village has a primary school, but no secondary school or health centre. It has no electricity, piped water or proper sewage disposal. In addition, there are areas of rainforest and hot desert where people find it extremely hard to make a living.

**FIGURE 4.10**

**FIGURE 4.11** *Wealthy area of Mexico City*

## 4D Other indicators of standard of living

Standard of living indicators can be divided into those that measure wealth (for example GNP per capita) and those that measure **quality of life** (for example population per doctor). There are several ways of measuring these.

### 1 Wealth

A wealthy country can afford to feed its people well and can build schools, hospitals and provide other services, so that everyone can enjoy a high standard of living.

Indicators of wealth are **average income per person**, **gross domestic product per capita (GDP per capita)** and **gross national product per capita (GNP per capita)**. GDP is the value of goods and services produced in a country in a year. GNP = GDP + the value of services earned abroad.

### 2 Food Intake

To have a high standard of living people must at least have enough food to eat. Many people eat food with enough calories but do not necessarily have a balanced diet. So, as well as the **number of calories per person per day**, the **amount of protein per person per day** is often used as an indicator of being well fed.

### 3 Education

Well-educated people can do skilled work, can think of new ideas, can become teachers, doctors, engineers or have other occupations which help to improve everyone's standard of living. The **percentage of people who are literate** does not always tell us exactly how well educated the people are. The **percentage of children at secondary school** is sometimes used instead.

### 4 Health

People cannot have a high standard of living if they are not healthy most of the time. The **population per doctor** does not tell us how many people are ill. Other useful indicators are **life expectancy** (how long the average person can expect to live) and **infant mortality** (the proportion of children who die before they are one year old).

### 5 Industrialisation

If most people work in factories and offices, many goods can be produced and sold. This increases the wealth of the country which can be used to improve the people's standard of living. However, the **percentage of people in agriculture** does not tell us how many goods are produced. The **amount of energy used per capita** is sometimes used instead.

## 4E Problems with indicators

1 If the population has not been counted properly, all the averages used are unreliable.
2 Averages can hide big differences. Some people may be well fed or wealthy while others may be very hungry or very poor.
3 One indicator of standard of living is not enough. Being able to read does not make up for being hungry. Being wealthy does not make up for being ill.

### >> EXTENSION QUESTIONS

*Look at the Extension Text.*
GDP per capita; number of calories per person per day; percentage of people who are literate; life expectancy; amount of energy used per capita

1 Which of the above are (a) indicators of wealth, and (b) indicators of quality of life?

2 What is meant by infant mortality?

3 Why is education a good indicator of standard of living?

4 Why can averages sometimes be misleading?

5 Why is it better to use more than one indicator to measure standard of living?

## Case Study of north and central America

**1** *Look at Figure 4.6*
  (a) Rank the countries of north and central America from 1–10 according to each indicator of standard of living (1 = highest; 10 = lowest).
  (b) Add the rankings for each country and work out an overall ranking.
  (c) Which of the ten countries are developed countries? Give reasons for your answer.

**2** Some people think that the five indicators in Figure 4.6 accurately show a country's standard of living. Others disagree.
  (a) Describe the different arguments that people would put forward.
  (b) Suggest alternative indicators that could be used.

**3** *Look at Figure 4.10.*
  Compare the standard of living in the cities and countryside of central America.

# World Population Growth

## Core text

### 5A The world's population

Figure 5.1 opposite shows the world's population growth over the last 300 years. The population is growing rapidly, more rapidly than ever before. It is doing so because there are a lot more births each year than deaths.

To find out why the population of the world is growing, we must find out why there are more births than deaths.

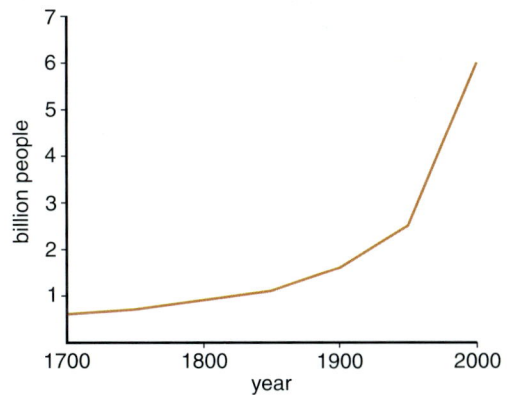

### 5B Birth rates and death rates

**Birth rate** = the number of births for every 1000 people each year.

Each year there are 21 births for every 1000 people in the world. This can be written as 21‰.

**Death rate** = the number of deaths for every 1000 people each year.

Each year there are 9 deaths for every 1000 people in the world. This can be written as 9‰.

**Natural increase in population** (the number of extra people) 5 birth rate – death rate.

The natural increase in the world = 21‰ − 9‰ = 15‰.

So for every 1000 people in the world, there are 12 more at the end of each year.

### 5C Differences in birth rates

The average birth rate in the world is 21‰, but some countries have a much higher birth rate than others (see Figure 5.2).

**Developed countries have a low birth rate.**
**Developing countries have a high birth rate.**

The reasons why birth rates vary are shown in Figure 5.3 on the next page.

**FIGURE 5.1** *Growth of the world's population*

| COUNTRY | BIRTH RATE (‰) |
|---|---|
| Argentina | 18 |
| Ethiopia | 44 |
| India | 24 |
| Liberia | 46 |
| Netherlands | 12 |
| United Kingdom | 11 |

**FIGURE 5.2** *Birth rate per country*

| COUNTRY | DEATH RATE (‰) |
|---|---|
| Afghanistan | 17 |
| Ethiopia | 18 |
| Japan | 9 |
| Sierra Leone | 19 |
| United Kingdom | 10 |
| Zambia | 22 |

**FIGURE 5.4** *Death rate per country*

**FIGURE 5.3** *Reasons for differences in birth rates*

## 5D Differences in death rates

The average death rate in the world is 9‰, but some countries have a higher death rate than others (see figure 5.4).

Developed countries have a low death rate. Developing countries have a higher death rate.

Figure 5.5 shows some of the reasons why the death rate is higher in developing countries.

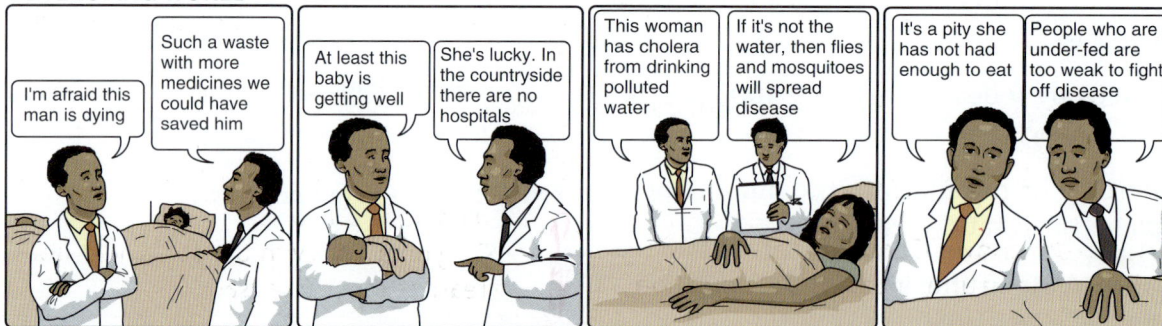

**FIGURE 5.5** *Reasons for differences in death rates*

## 5E Differences in population growth

Developed countries are growing slowly in population because there are only slightly more births each year than deaths. Developing countries are growing rapidly in population because there are many births each year but deaths are only slightly higher than in developed countries.

In developed countries, birth and death rates do not change much from year to year. In developing countries, birth rates are falling but death rates are falling even faster.

### CORE QUESTIONS

**1** *Look at Figure 5.1.*
In what ways is the world's population changing?

**2** *Look at 5B.*
Spain has a birth rate of 9‰. Does this mean:
(a) 9 people are born each year
(b) for every 1000 people, there are 9 born each year
(c) 900 are born each year?

| COUNTRY | BIRTH RATE (‰) | DEATH RATE (‰) | NATURAL INCREASE (‰) |
|---------|---------------|----------------|----------------------|
| Afghanistan | 41 | 17 | 24 |
| Australia | 13 | 7 | |
| Ethiopia | 44 | 18 | |
| Japan | 10 | 9 | |
| Liberia | 46 | 16 | |
| Italy | 9 | 10 | |

**3** Copy the table above. Complete the table by working out the natural increase in population for each country.

**4** *Look at Figure 5.3.*
Give reasons why people in developing countries often have many children.

**5** Give reasons why people in developed countries often have few children.

**6** *Look at Figure 5.5.*
Why do people live longer in developed countries than in developing countries?

**7** *Look at 5E.*
Which countries are growing more quickly, developed or developing countries?

### FOUNDATION QUESTIONS

## Case Study of India

**1** *Look at Figure 5.9.*
In which year did India have:
(a) its highest birth rate
(b) its lowest birth rate
(c) its highest death rate
(d) its lowest death rate
(e) its highest natural increase
(f) its lowest natural increase?

**2** *Look at Figure 5.10.*
How has India's population changed since 1950?

**3** *Look at Figure 5.11.*
In 1950 there were no pensions or sickness benefits in India. Do you think this affected the number of children people had? Give reasons for your answer.

**4** Suggest two reasons why Indian couples had large families in the 1950s.

**5** There are now fewer children dying in India. Do you think this has affected the birth rate? Give reasons for your answer.

**6** Suggest one reason why birth rates in India have dropped recently.

**7** *Look at Figure 5.12.*
(a) To what age could people expect to live in the 1950s?
(b) Suggest two reasons why people did not live very long.'

**8** 'People live longer in India now because there is more food available.'
Do you agree with the statement above? Give reasons for your answer.

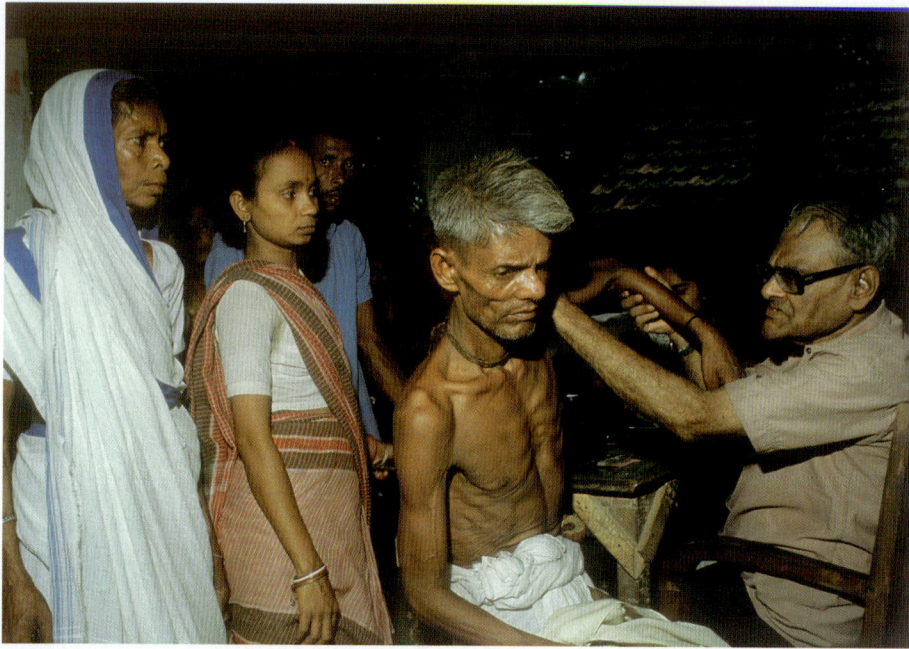

**FIGURE 5.6** *Health care in India*

## Case Study of India

**1** *Look at Figure 5.9.*
Compare the changes in birth and death rates since 1950.

**2** *Look at Figure 5.10.*
Describe the changes in India's population since 1950.

**3** *Look at Figure 5.11.*
Today fewer people in India live in the countryside. Do you think this explains why the birth rate is falling? Give reasons for your answer.

**4** What, do you think, is the main reason why birth rates have dropped since 1960? Give reasons for your answer.

**5** *Look at Figure 5.12.*
Suggest reasons why the death rate in India was high in 1950.

**6** Why do you think the death rates are now lower?

# RESOURCES

## Case Study of India

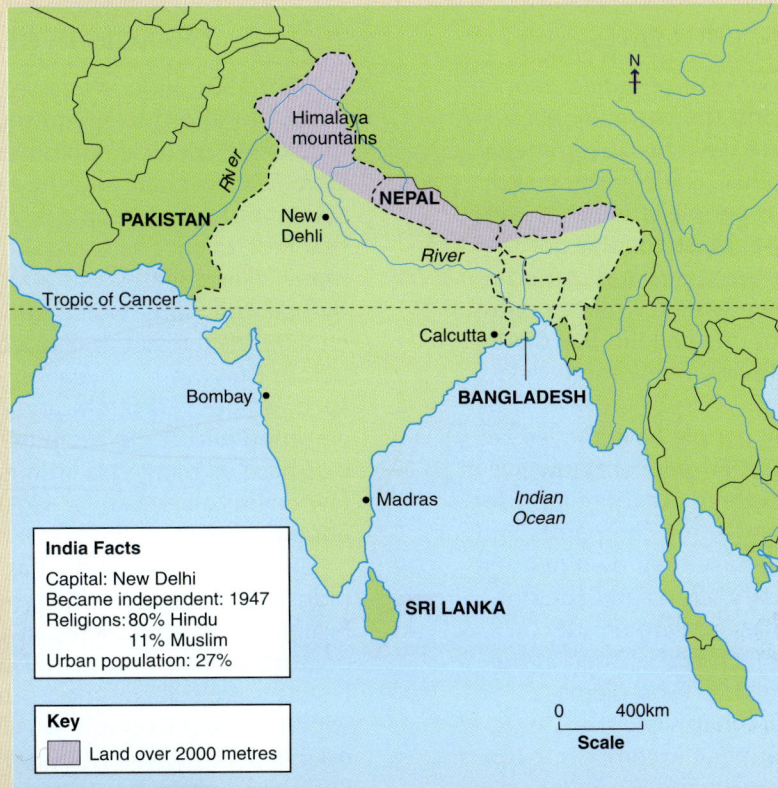

**FIGURE 5.7** India

## INTRODUCTION

India is a large country in south Asia. It has the second biggest population in the world (behind China) and it is growing rapidly. Until recently, it grew rapidly because there was a very high birth rate. Indian couples wanted to have many children. Today, the birth rate is lower but so is the death rate, which means the population continues to grow as fast as ever before. Figure 5.10 shows how the population has grown in recent years, and Figure 5.9 shows the changes in birth rates and death rates.

**FIGURE 5.8**

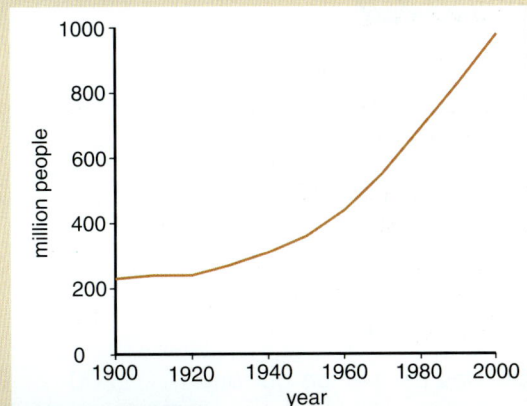

**FIGURE 5.10** India's population growth

| | BIRTH RATE (‰) | DEATH RATE (‰) | NATURAL INCREASE (‰) |
|---|---|---|---|
| 1950 | 42 | 27 | 15 |
| 1960 | 44 | 22 | 22 |
| 1970 | 42 | 19 | 23 |
| 1980 | 35 | 15 | 20 |
| 1990 | 32 | 14 | 18 |
| 2000 | 25 | 10 | 15 |

**FIGURE 5.9**

# RESOURCES

## CHANGES IN BIRTH RATES

In the 1950s, about 80 per cent of Indian people lived in the countryside and the vast majority were farmers. At that time there were no government pensions or sickness benefits. The country could not afford to look after its people when they were old or were too ill to work. Most couples got married when they were teenagers and started families immediately. There were few birth control measures available. About one child in five died before it was one year old and many more died before adulthood.

Since the 1950s, many people have moved to towns and cities and about 70 per cent now live in the countryside. In towns there are more jobs, for both men and women, and some companies and government departments have their own pension schemes. Couples still marry at an early age but now nearly one half of married women use contraception, which is much higher than in the 1950s. The number of children who die has dropped by more than one half since the 1950s. The government has also made it compulsory for all children to attend primary school.

**FIGURE 5.11**

## CHANGES IN DEATH RATES

In the 1950s, India was a very poor country and could not afford to spend much money on medicines, medical equipment and doctors. There were few hospitals and they were only found in towns and cities. People living in the countryside had little chance of using a hospital. There were many killer diseases and a lot of these were spread by polluted water which people drank, for example cholera. As a result, most people in India died before they were 40 years of age. Natural disasters, such as floods, drought and cyclones, were common and could ruin farmers' harvests. Farming was hard as there was little equipment and very few chemicals available. Most people did not get enough to eat.

Since the 1950s, India has become more industrialised and is a little wealthier. Its health facilities have improved considerably, with more doctors, nurses, hospitals and health centres. Diseases are still common but some have been wiped out, for example smallpox, and people can now expect to live until they are 63 years old. Food production has increased greatly and generally people are much better fed.

**FIGURE 5.12**

|  | 1950 | 2000 |
|---|---|---|
| GNP/person | £70 | £350 |
| population per doctor | 6000 | 2400 |
| life expectancy | 38 | 63 |
| calories per person per day | 1700 | 2400 |
| infant mortality | 190% | 61% |
| access to safe water | 45% | 88% |

**FIGURE 5.13**

**5F** **Model of population change**

Every country's birth and death rates change over time. The graph below shows the typical changes in birth rates and death rates in a country. It is called a **model of population change**.

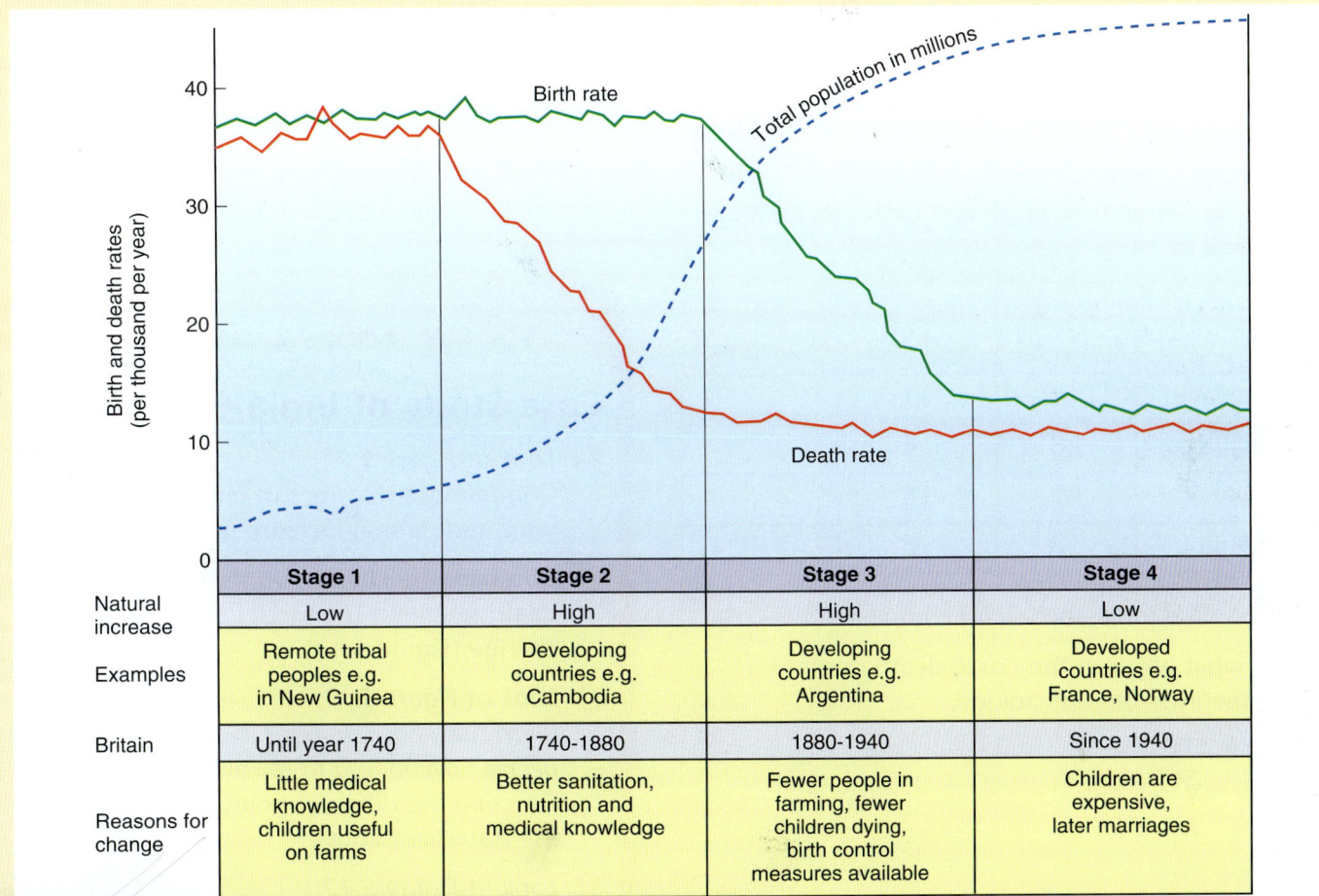

| | Stage 1 | Stage 2 | Stage 3 | Stage 4 |
|---|---|---|---|---|
| Natural increase | Low | High | High | Low |
| Examples | Remote tribal peoples e.g. in New Guinea | Developing countries e.g. Cambodia | Developing countries e.g. Argentina | Developed countries e.g. France, Norway |
| Britain | Until year 1740 | 1740-1880 | 1880-1940 | Since 1940 |
| Reasons for change | Little medical knowledge, children useful on farms | Better sanitation, nutrition and medical knowledge | Fewer people in farming, fewer children dying, birth control measures available | Children are expensive, later marriages |

**FIGURE 5.14** *Model of population change*

**5G** **Problems in using birth rates and death rates**

The **crude birth rate** is the number of children born for every 1000 people. This can be misleading as it depends upon how many young women there are in the population. Crude birth rates are higher in developing countries partly because there is a greater proportion of women in the 15–30 age group.

Equally the crude death rate (number of deaths for every 1000 people) can be misleading as it depends upon the number of old people in the population. In many developed countries the crude death rate is rising, not because people are dying younger, but because there is a higher percentage of old people in the population.

# EXTENSION TEXT

*Look at the Extension Text.*

1. What is meant by a model of population change?

2. Explain why there is a high birth rate in Stage 1 of the model.

3. Explain why there is a lower death rate in Stage 2.

4. Explain why there is a lower birth rate in Stages 3 and 4.

| COUNTRY | BIRTH RATE (%) | DEATH RATE (%) | STAGE |
|---------|----------------|----------------|-------|
| Sweden | 10 | 11 | |
| Philippines | 28 | 6 | |
| Nigeria | 41 | 14 | |
| China | 13 | 6 | |

5. Copy the table above and complete it.

6. In what way can the crude death rate sometimes be misleading?

## CREDIT QUESTIONS

### Case Study of India

1. *Look at Figure 5.9.*
   Compare the changes in birth rates, death rates and natural increase in India since 1950.

2. Compare the changes in India's birth rate and death rate with the model of population change (Figure 5.14).

3. *Look at Figure 5.11.*
   *'Birth rates have dropped in India because many people have moved to towns.'*
   Describe the different points of view towards the statement above.

4. *Look at Figure 5.12.*
   Suggest why the death rate was high in India in the 1950s.

5. What is the most important reason why the death rate has fallen since the 1950s? Give reasons for your answer.

# 6

# The Effects of Population Change

## Core text

### 6A Differences in population growth

In the last unit we found out that developing countries are growing rapidly. This brings many problems, as well as some benefits to these countries. Developed countries are only growing slowly and this, too, brings both advantages and disadvantages. In this unit, we shall study the different effects of rapid and slow population growth.

### 6B Dependent and active population

In any country, the **dependent population** are those people who are too old (65 years old or over) or too young (15 years old or younger) to work. The **active population** are those people aged 15–65, who make up the country's workforce and provide for those who are too old or too young to work.

The active population are the most useful age group, not only because they produce wealth, but also because they pay taxes which gives the government money to pay for schools, hospitals, roads, houses etc. They also provide the people to go into the armed forces and defend the country.

The dependent population generally does not produce any wealth and they also need looking after, which costs the country money. The different ways in which both young and old people have to be provided for are shown in Figures 6.1 and 6.2.

**FIGURE 6.1** *Money spent on young people*

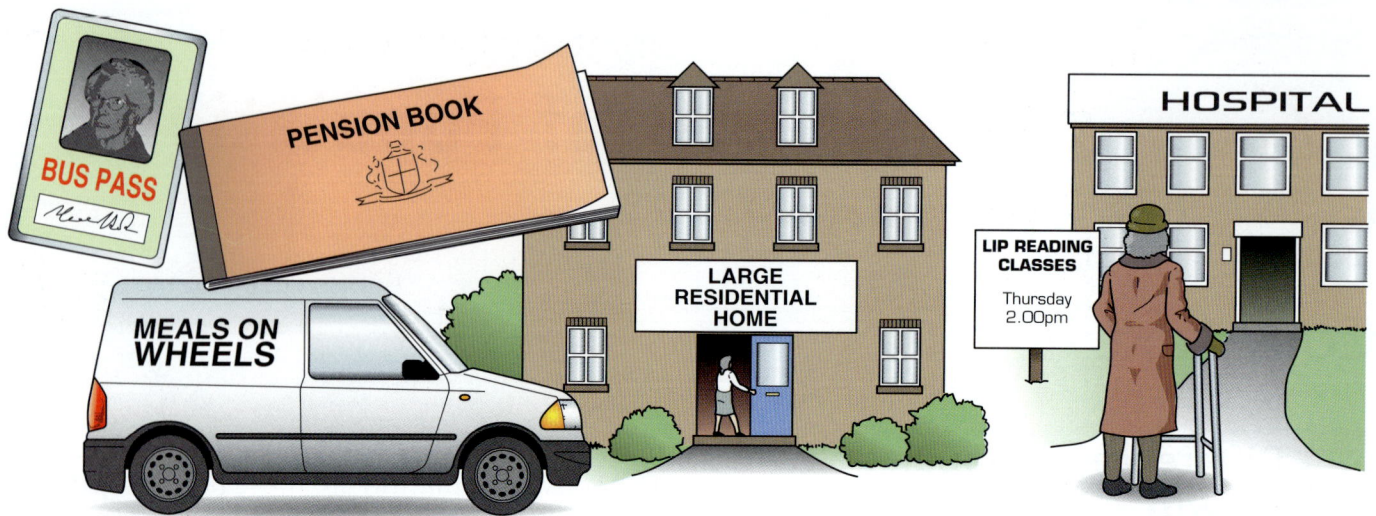

**FIGURE 6.2** *Money spent on senior citizens*

## 6C Effects of a rapidly growing population

If a country has a rapidly growing poplation it has a lot of young people, relatively few adults and few old people. This affects the country in various ways:

1 The many young people have to be provided with health care, schools etc.
2 There are fewer adults to produce the country's wealth.
3 It saves money by not having many old people to look after.

In the countryside, it means that:
1 Farms become smaller, so each family grows less food.
2 More trees have to be cut down, so the soil becomes poorer.
3 There are not enough jobs for everyone on the land, so people leave for the cities.

In the cities it means that:
1 There are not enough jobs, so many people are unemployed.
2 There are not enough houses, so people are either homeless or build their own rough shacks.
3 There are not enough schools and hospitals for everyone to use.

## 6D Effects of a slowly growing population

If a country has a slowly growing population, it has few young people, a decreasing number of adults and many old people. This affects the country in various ways:

1 It saves money by not having many children to provide for.
2 Fewer children mean fewer adults in the future to produce the country's wealth and pay taxes.
3 The many old people need pensions, health care and other social services.

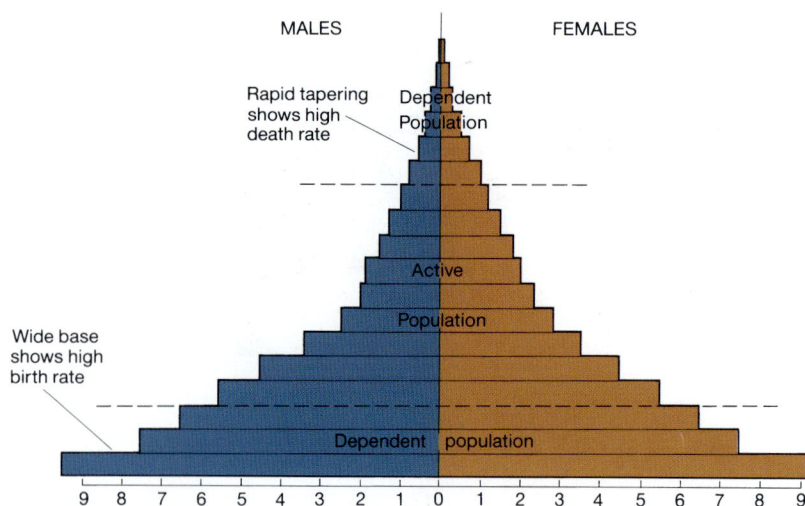

**FIGURE 6.3** *Population pyramid for a rapidly growing population*

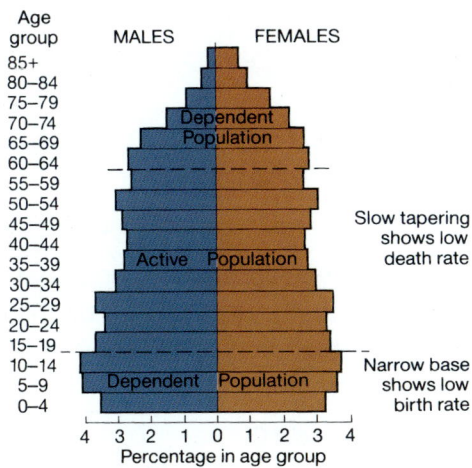

**FIGURE 6.4** *Population pyramid of a slowly growing population*

## 6E Solving the problems of a rapidly growing population

**Solution 1 – Reduce Birth Rates**

Many approaches have been used to try and reduce birth rates:

1 Family planning clinics are set up to give advice and to distribute contraceptives.
2 Couples are fined if they have large families.
3 Couples are rewarded if they have small families, for example free education.
4 Abortions and sterilisations are made easier.
5 Couples are not allowed to marry until they are older.
6 Couples are only allowed to have children several years apart.

**Solution 2 – Increase the Country's Wealth**

Instead of, or as well as, reducing birth rates, a country can try to increase its wealth. Then, even if there are more people, there is more wealth to be shared between them and the people should be no poorer. Possible approaches include:

1 Finding and exploiting more natural resources (for example, coal, oil, iron ore).
2 Using farmland better (for example fertilisers, irrigation).
3 Making more farmland (for example clearing forests).
4 Setting up manufacturing and tourist industries.

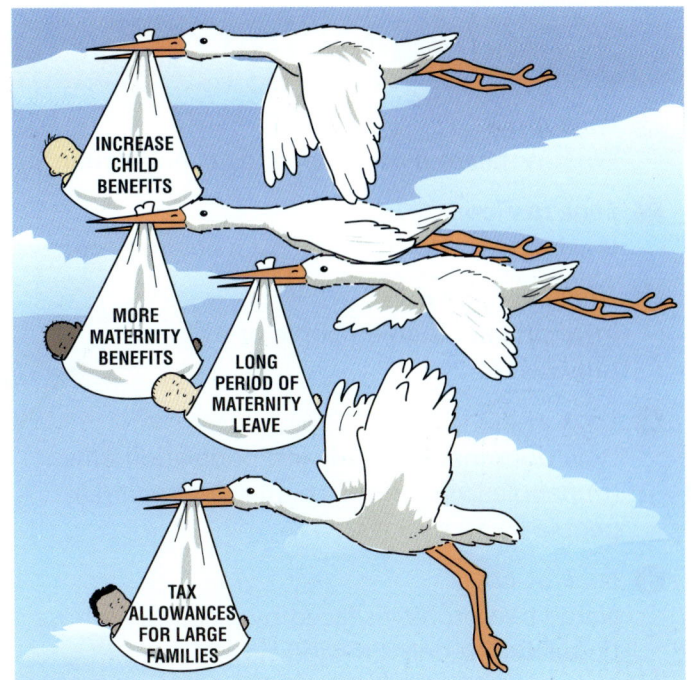

**FIGURE 6.5** *Methods of increasing birth rates*

## 6F Solving the problems of a slowly growing population

**Solution 1 – Increase Birth Rates**

Different methods of increasing birth rates are shown in Figure 6.5.

**Solution 2 – Increase the Workforce**

This might be done by:

1 Persuading more women to take jobs (for example by providing crèches in workplaces, giving women equal pay and conditions).
2 Raising the age of retirement.
3 Increasing the number of immigrants.

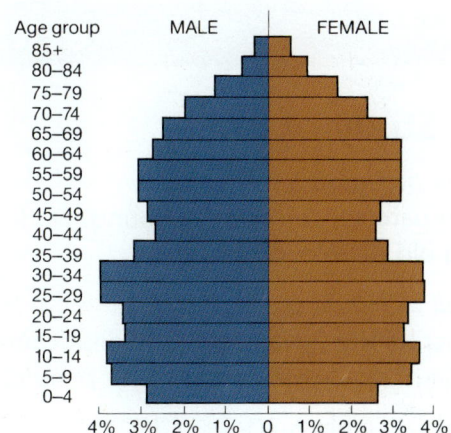

**FIGURE 6.6** *Population pyramid*

1 *Look at 6B.*
What is meant by the 'dependent population'?

2 *Look at Figure 6.1.*
In what ways do children cost a country money?

3 *Look at Figure 6.2.*
In what ways do old people need to be provided for?

4 *Look at 6C.*
How does a rapidly growing population affect (a) people living in the countryside, and (b) people living in cities?

5 *Look at 6D.*
Name two problems faced by a country whose population is only growing slowly.

6 *Look at Figure 6.6.*
Does the population pyramid show a high or low *birth* rate? Give a reason for your answer.

7 Does the population pyramid in Figure 6.6 show a high or low *death* rate? Give a reason for your answer.

8 *Look at 6E.*
Describe three ways in which countries can reduce their birth rates.

9 *Look at 6F.*
Name two ways in which countries can increase their birth rates.

10 Describe how countries can increase their workforce.

## Case Study of India

1 *Look at Figures 6.8 and 6.9.*
Does India have a lot of old peple or very few?

2 (a) How can you tell from the population pyramid that India has a lot of children?
(b) Do you think that this is good or bad for the country?

3 *Look at Figure 6.10.*
(a) What is happening to the size of farms in India?
(b) What problems does this bring?

4 *Look at Figure 6.11.*
Why do Indian cities, such as Mumbai, formally called Bombay, not have enough jobs or houses for their people?

5 *Look at Figure 6.12.*
How has India managed to feed its growing population?

6 *Look at Figure 6.13.*
(a) Describe the different ways in which India has tried to reduce its birth rate?
(b) Which has been the most successful? Give reasons for your answer.

## Case Study of India

1 *Look at Figure 6.8.*
Compare the numbers of young, working age and old people in India.

2 *Look at Figure 6.9.*
Describe the advantages and disadvantages to India of having so many young people.

3 *Look at Figure 6.10.*
Describe the effects of a growing population on the countryside areas.

4 *Look at Figure 6.11.*
What evidence is there that Indian cities, such as Mumbai, cannot cope with their increasing populations?

5 *Look at Figure 6.13.*
Describe the measures India has introduced to reduce birth rates.

6 (a) In which states have birth rates fallen most?
(b) Suggest some reasons for this.

## Case Study of India

### INTRODUCTION

We found out in the last unit that India's population is growing rapidly. Unless it slows down, there will be 1600 million people living in the country by the year 2040 and it will have more people than any other country in the world. The Indian government is aware of the problems that this will cause and has been trying to control population growth since 1952. Some of the problems and attempted solutions are studied here.

**FIGURE 6.7**

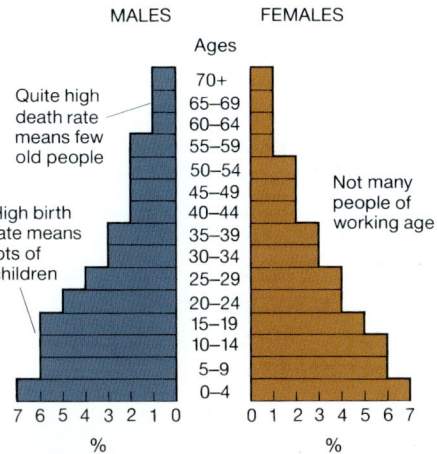

MALES    FEMALES

Ages

Quite high death rate means few old people

High birth rate means lots of children

Not many people of working age

70+
65–69
60–64
55–59
50–54
45–49
40–44
35–39
30–34
25–29
20–24
15–19
10–14
5–9
0–4

7 6 5 4 3 2 1 0    0 1 2 3 4 5 6 7

%    %

**FIGURE 6.8**  *India's population structure*

### PROBLEMS IN THE COUNTRYSIDE

As the number of people living in the countryside has increased, the farms have become smaller and smaller. By the 1990s, over half of all farms were under one hectare in size (1 hectare is the size of a football pitch). Farms have now become so small that many farmers only have enough land to feed themselves and their families. They have no extra land to grow food for sale. As the population grows, these farms cannot become any smaller. Some people will have to leave the countryside to find work in the cities.

**FIGURE 6.10**

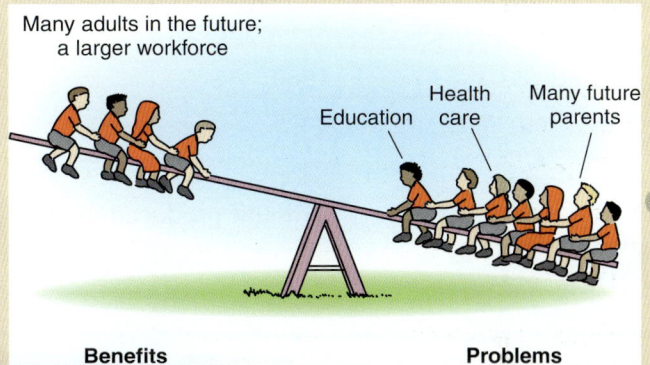

Many adults in the future; a larger workforce

Education    Health care    Many future parents

Benefits    Problems

**FIGURE 6.9**  *The benefits and problems of many children*

### PROBLEMS IN THE CITIES

Cities in India cannot cope with all the extra people. In Mumbai, India's biggest city, only 20 per cent of the people have full time jobs and only 30 per cent have proper housing. Over 100 000 people have no home at all. The authorities cannot supply every home with electricity and water and, even for those people lucky enough to have these, the supply is very unreliable. As the population continues to grow congestion and air pollution become worse.

**FIGURE 6.11**

# RESOURCES

## DEVELOPING INDIA'S WEALTH

One way of solving the problem of India's rapidly growing population is to ensure that the country produces more food and more industrial goods and so makes more money. In this way, although there will be more people in India, there will also be more wealth, so everyone can enjoy a higher standard of living. Since the 1950s, India has tried different ways of increasing its wealth.

**In the 1950s**, it began to exploit its minerals, especially coal and iron ore, and started up many factories such as steel, engineering and chemical industries. India has now become the twelfth most industrialised country in the world.

|  | 1950 | 1999 |
|---|---|---|
| Workers employed in industry | 2% | 17% |
| Exports which are manufactured goods | 54% | 79% |
| Total wealth produced by industry | 15% | 30% |

**In the 1960s**, India began to find ways of increasing the amount of food it grew. Farmers were given cheap loans to buy better seeds, fertilisers and pesticides. More irrigation schemes were started, which increased farmers' yields and gave them two harvests a year. As a result of these measures, although the population has grown, food production has grown even faster.

|  | 1960 | 2002 |
|---|---|---|
| Rice | 350 | 1200 |
| Wheat | 110 | 720 |
| Sugar cane | 1100 | 2790 |

(amount produced in 100 000 tonnes)

**FIGURE 6.12**

**FIGURE 6.14** Birth control poster in India

## REDUCING INDIA'S BIRTH RATES

Although India successfully increased its wealth in the 1950s and 1960s, its population continued to grow rapidly and the government could not be sure that it would always be able to provide for an ever increasing population. It decided it must reduce birth rates and has since tried several approaches.

**In the 1970s** men and women were given free gifts, such as transistor radios, if they agreed to be sterilised. For a short time, compulsory sterilisation was brought in and ten million people were sterilised in two years. However, this policy met with such disapproval that it was stopped.

**In the 1980s and 1990s**, India has had more success in reducing birth rates. They first began to fall in states such as Goa, Kerala and Tamil Nadu where there is higher literacy (especially among females), low infant mortality, a higher age of marriage and greater use of contraceptives. So, to try and reduce birth rates in the rest of the country, the government has set up many family planning clinics, giving people advice on birth control. It has also raised the legal age of marriage to 18 for women and 21 for men and it is trying hard to improve education and health throughout the country.

**FIGURE 6.13**

## 6G Overpopulation and underpopulation

If a country is **overpopulated**, it means there are too many people for the resources available. So the people's standard of living is low.

If a country is **underpopulated**, it means there are too few people to develop the resources the country possesses. So the people's standard of living will also be low.

If a country has its **optimum population**, it has just enough people to develop its resources fully so that everyone enjoys a high standard of living. The optimum population will not be the same in every country. It depends on the resources each country possesses.

An overpopulated country can solve its problem by reducing its population and/or improving ways of using its resources, for example developing tourism, farming. An underpopulated country can solve its problem by increasing its population and/or using its resources more efficiently, for example using machines instead of workers.

## 6H Dependency ratios

The rate at which a population grows affects the population structure. Every country would like a population in which there is a high percentage of active people and a low percentage of dependent people.

The **dependency ratio** shows how 'active' a country's population is:

$$\text{dependency ratio} = \frac{\%\ \text{dependent population}}{\%\ \text{active population}}$$

The lower the dependency ratio, the more 'active' the population is.

| COUNTRY | DEPENDENT POPULATION | ACTIVE POPULATION | DEPENDENCY RATIO |
|---------|----------------------|-------------------|------------------|
| UK | 39% | 61% | 0.64 |
| USA | 38% | 62% | 0.62 |
| Norway | 40% | 60% | 0.67 |
| India | 44% | 56% | 0.79 |
| Pakistan | 50% | 50% | 1.00 |
| Nigeria | 51% | 49% | 1.04 |

## >> EXTENSION QUESTIONS

*Look at the Extension Text.*

1. Why does a country have a low standard of living if it is (a) overpopulated, and (b) underpopulated?

2. Why is it not possible to work out whether a country is overpopulated from its population density?

3. How do the dependency ratios of developed and developing countries differ?

4. How is a low dependency ratio of benefit to a country?

## Case Study of India

1. *Look at Figure 6.8.*
   How can you tell that India's population is growing rapidly?

2. *Look at Figures 6.8 and 6.9.*
   Do you think India's dependency ratio is helpful or harmful to the country's economy? Give reasons for your answer.

3. *Look at Figures 6.10 and 6.11.*
   To which areas has India's rapid population growth brought greater problems, the cities or the countryside? Give reasons for your answer.

4. *Look at Figure 6.12.*
   India has tried many approaches to its population problem. Describe the different points of view people would have had towards its approach in the 1950s and 1960s.

5. *Look at Figure 6.13.*
   Suggest reasons why India's birth rate has fallen faster in the 1980s and 1990s than in the 1970s.

# 7 Migration Within the Developing World

## Core text

### 7A Migration

The number of people in a country or region depends upon how many births and deaths there are. It also depends upon how many people move into the region (**immigrants**) and how many people move away from the region (**emigrants**).

### 7B Migration from the countryside to cities

In developing countries, millions of people migrate each year from the countryside (**rural areas**) to towns and cities (**urban areas**). Sometimes they move because of the problems in

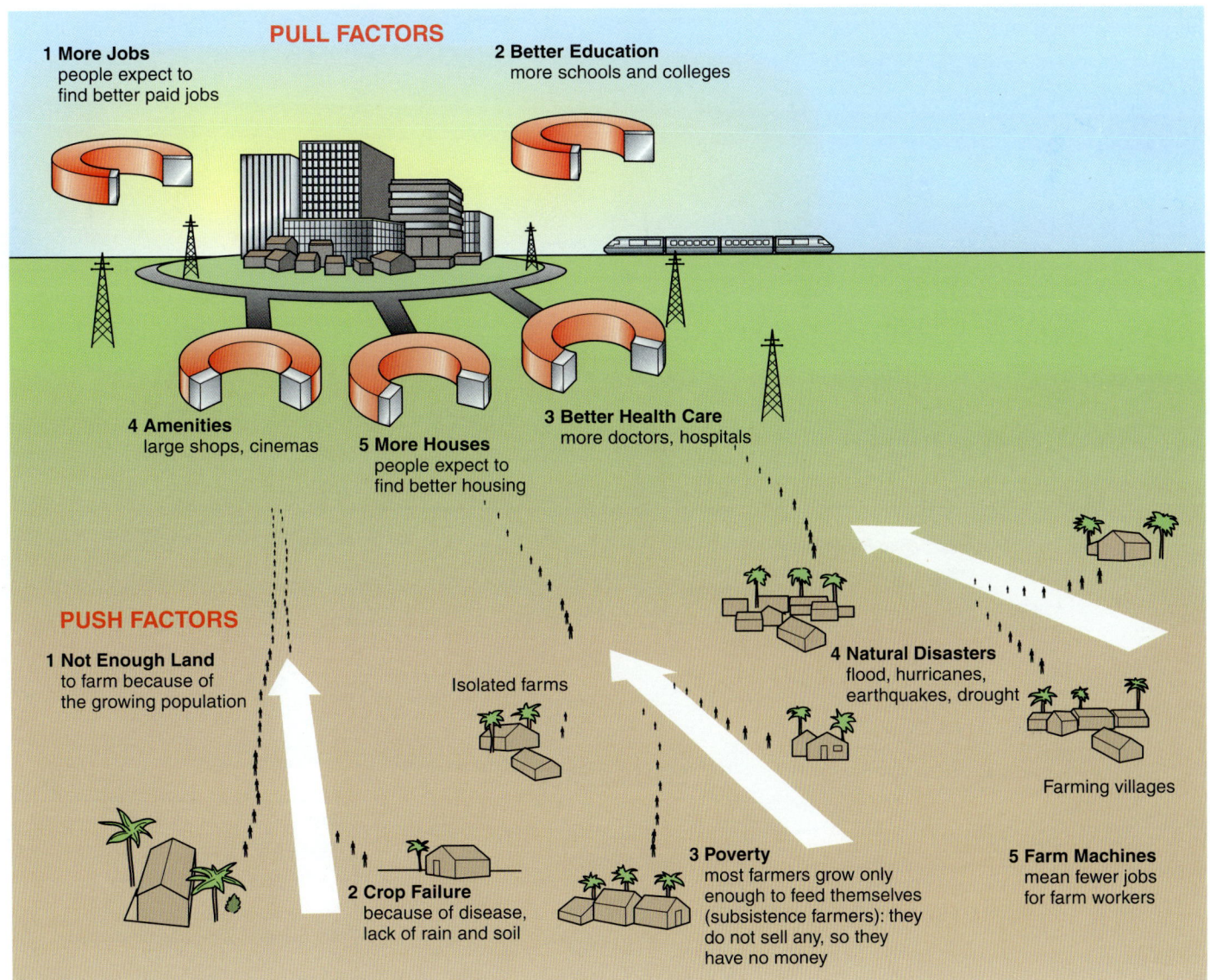

**PULL FACTORS**

**1 More Jobs**
people expect to find better paid jobs

**2 Better Education**
more schools and colleges

**4 Amenities**
large shops, cinemas

**5 More Houses**
people expect to find better housing

**3 Better Health Care**
more doctors, hospitals

Isolated farms

**4 Natural Disasters**
flood, hurricanes, earthquakes, drought

Farming villages

**PUSH FACTORS**

**1 Not Enough Land**
to farm because of the growing population

**2 Crop Failure**
because of disease, lack of rain and soil

**3 Poverty**
most farmers grow only enough to feed themselves (subsistence farmers): they do not sell any, so they have no money

**5 Farm Machines**
mean fewer jobs for farm workers

**FIGURE 7.1** *Reasons for migrating to developing world cities*

the countryside. These are called **push factors**. Sometimes they move because of the attractions of the cities. These are called **pull factors**. The main push and pull factors are shown in Figure 7.1.

There are however some advantages. The countryside is less overcrowded and the migrants often send back money to their families still living in the countryside.

## 7C Effects of migration on the countryside

Figure 7.2 on the next page shows a population pyramid from a countryside area in a developing country. There are few young adults in the population because many migrate to the cities. So the countryside loses its most active population, the ones who can do the most work and have the most ideas. The farming has to be done by middle aged and old people, and especially by women.

## 7D Urbanisation

Cities in the developing world are growing very rapidly. This is called urbanisation. In 1950, 250 million people lived in developing world cities, but by the year 2000 this had risen to 2000 million. These cities are growing rapidly because of the large number of people who migrate there from the countryside. Many now have over one million people and are called **millionaire cities**. Figure 7.3 shows the 20 largest cities in the world. Of these 20, 16 are in the developing world.

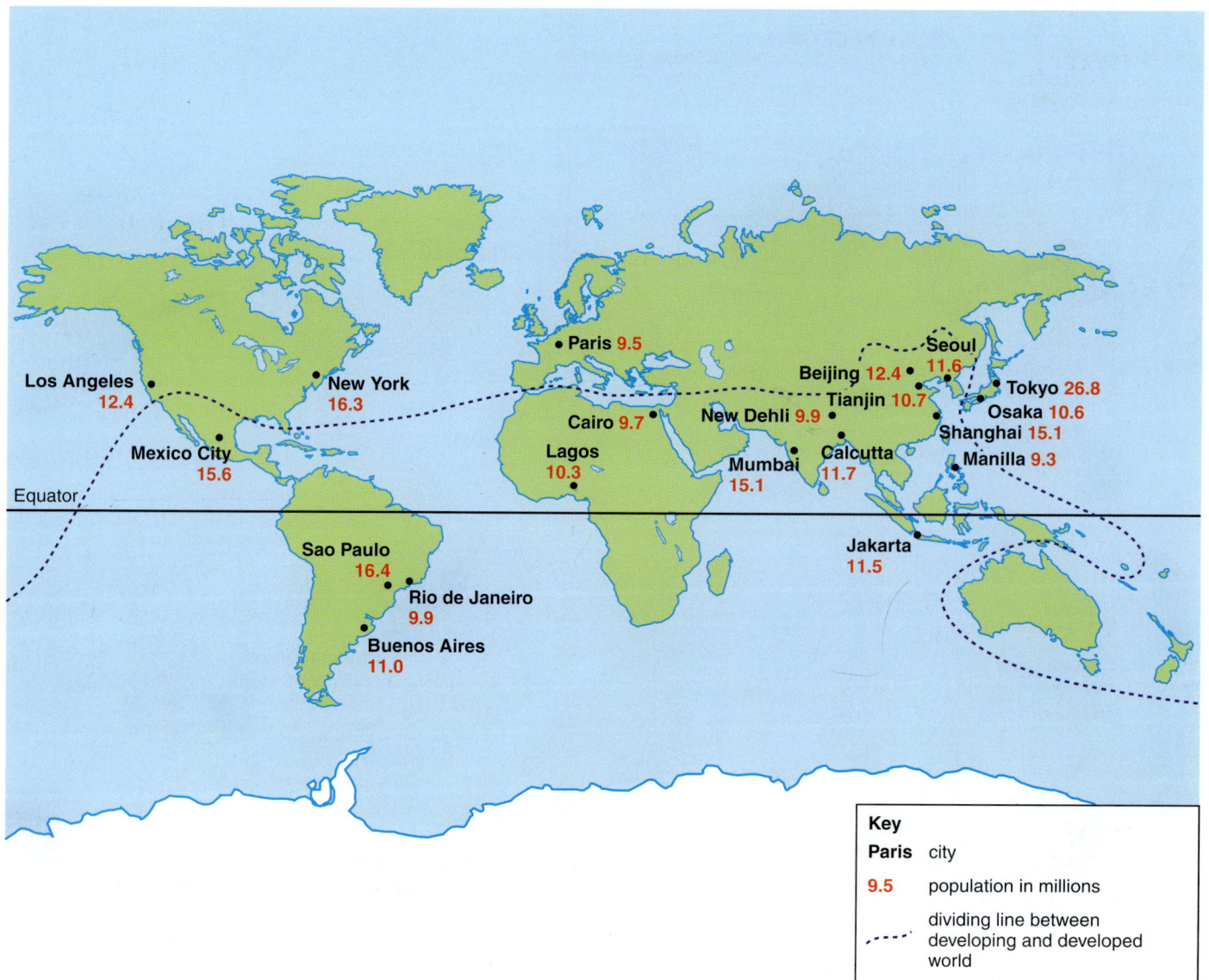

**Key**

**Paris** city

9.5 population in millions

- - - - dividing line between developing and developed world

Los Angeles 12.4
New York 16.3
Mexico City 15.6
Equator
Sao Paulo 16.4
Rio de Janeiro 9.9
Buenos Aires 11.0
Paris 9.5
Cairo 9.7
Lagos 10.3
Beijing 12.4
Seoul 11.6
Tianjin 10.7
Tokyo 26.8
Osaka 10.6
Shanghai 15.1
New Dehli 9.9
Mumbai 15.1
Calcutta 11.7
Manilla 9.3
Jakarta 11.5

**FIGURE 7.3** *Top 20 world cities*

## 7E Effects of migration on the cities

Cities in developing countries cannot cope with so many immigrants. There are not enough houses for everyone so people have to build makeshift houses on waste land. These squatter camps are called **shanty towns**, but they are not separate towns. They are inside the cities. About 40 per cent of people living in cities live in shanty towns and some of their problems are shown in Figure 7.4. As well as housing problems, there are not enough full time jobs for all the people. Most people, adults and children, have **informal jobs**, for example running errands, selling fruit, collecting rubbish. This work is irregular and unreliable and barely gives people enough money on which to survive. There are also severe environmental problems in the cities, such as air pollution and a lack of clean water, sewage treatment and rubbish disposal.

**FIGURE 7.2** *A population pyramid for a rural area in the Developing World*

### CORE QUESTIONS

1. Look at 7A.
   What is the difference between immigrants and emigrants?

2. Look at 7B.
   What is (a) a rural area, and (b) an urban area?

3. Look at Figure 7.1.
   Describe three push factors which cause people to move from the countryside.

4. Describe three pull factors which attract people to the cities.

5. Look at 7C.
   (a) Which age group is most likely to emigrate from the countryside?
   (b) In what ways does this affect the countryside?

6. Look at 7D.
   What is meant by urbanisation?

7. Look at Figure 7.3.
   Name the three biggest cities in the world.

8. Look at 7E.
   What are informal jobs?

9. What is a shanty town?

10. Look at Figure 7.4.
    Describe some of the problems of shanty towns.

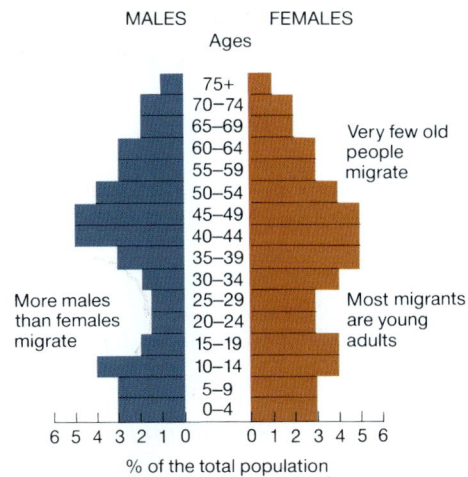

### FOUNDATION QUESTIONS

## Case Study of Peru

1. Look at Figure 7.8.
   Give two reasons why people might leave the Andes mountains.

2. Which area of Peru do you think has the greater problems, the coast or the rainforest? Give reasons for your answer.

3. Look at Figure 7.8 and 7.9.
   Describe the living conditions in the Andes mountains.

4. Why are farms in the mountains becoming smaller?

5. Look at Figure 7.12.
   Give two reasons why people are moving to Lima.

6. Look at Figure 7.13.
   Do you think that the large numbers of people migrating to the cities helps those people who stay in the countryside? Give reasons for your answer.

7. Look at Figure 7.14.
   Why is there so much disease in Lima's shanty towns?

8. Look at Figure 7.5.
   What message is the cartoon trying to get across?

## Case Study of Peru

**1** *Look at Figure 7.6.*
Describe the main migration movements in Peru.

**2** *Look at Figure 7.8.*
Describe some of the physical problems of living in the Peru countryside.

**3** *Look at Figures 7.9, 7.10 and 7.11.*
Suggest why the amount of food (calories) per person in Peru has gone down since 1950.

**4** *Look at Figure 7.12.*
Describe the attractions of Lima to people living in the countryside.

**5** *Look at Figure 7.13.*
Many people are now leaving the countryside. Describe one benefit and one problem this brings to the people left in the countryside.

**6** *Look at Figure 7.14.*
Suggest why disease is so common in the shanty towns of Lima.

**7** *Look at Figure 7.5.*
(a) What message is the cartoon trying to get across?
(b) Do you agree with the message? Give reasons for your answer.

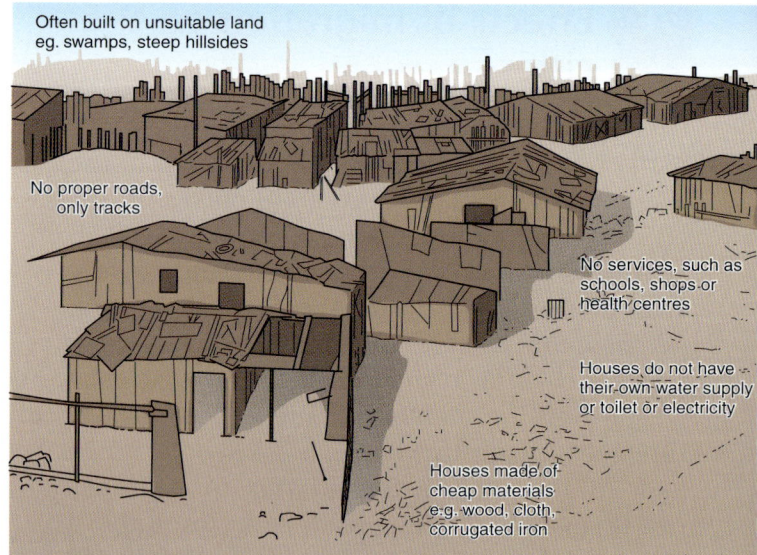

Often built on unsuitable land eg. swamps, steep hillsides

No proper roads, only tracks

No services, such as schools, shops or health centres

Houses do not have their own water supply or toilet or electricity

Houses made of cheap materials e.g. wood, cloth, corrugated iron

**FIGURE 7.4** *Problems in a shanty town*

**FIGURE 7.5**

## Case Study of Peru

### INTRODUCTION

Peru is a developing country on the west coast of South America. Nearly one third of its 27 million people live in the capital, Lima. Lima is ten times bigger than any other city in Peru and it is to here that many people in the countryside migrate.

Unlike other parts of the world, more females migrate than males. They move partly because of the difficulties of living in the countryside and partly because of the attractions of Lima. When they migrate they affect the countryside that they leave and the city to which they move.

**FIGURE 7.6**

**FIGURE 7.7** Peru within South America

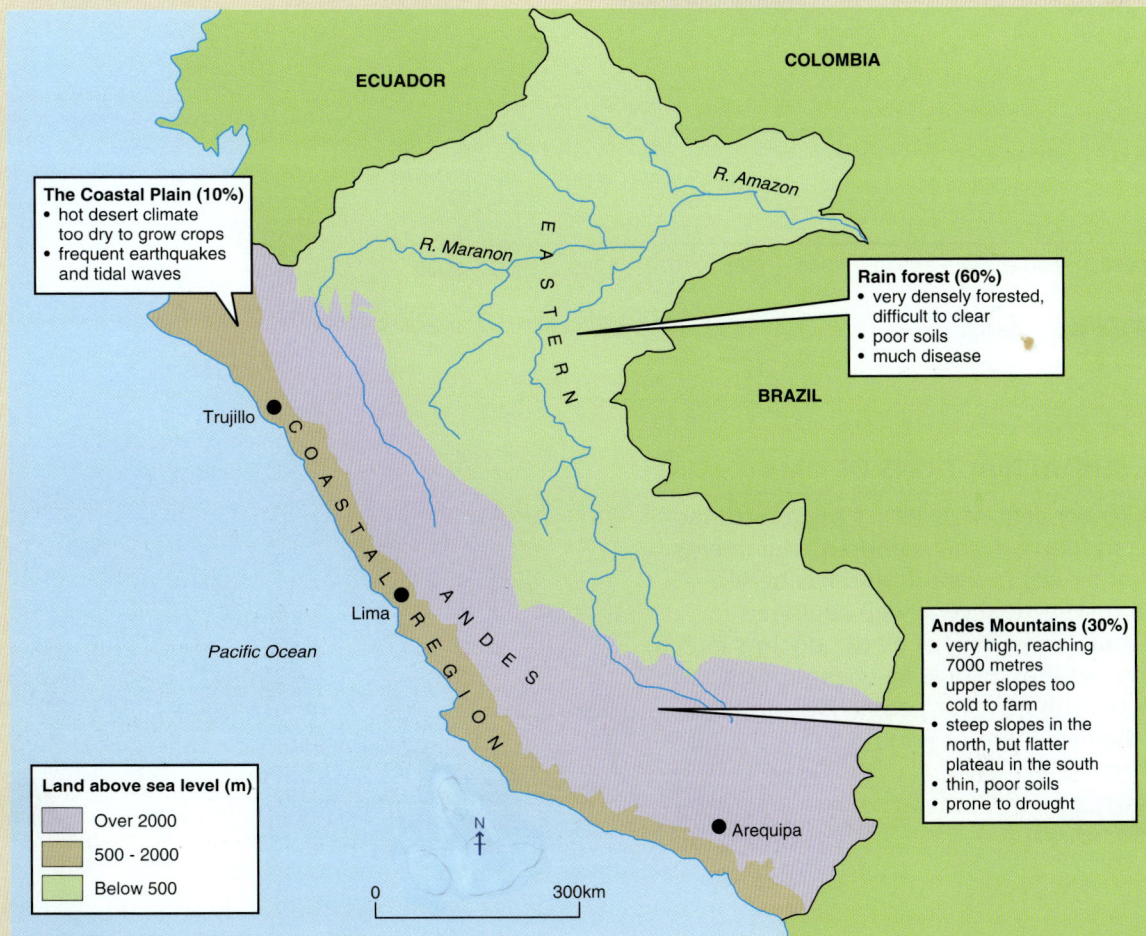

**The Coastal Plain (10%)**
- hot desert climate too dry to grow crops
- frequent earthquakes and tidal waves

**Rain forest (60%)**
- very densely forested, difficult to clear
- poor soils
- much disease

**Andes Mountains (30%)**
- very high, reaching 7000 metres
- upper slopes too cold to farm
- steep slopes in the north, but flatter plateau in the south
- thin, poor soils
- prone to drought

**Land above sea level (m)**
- Over 2000
- 500 - 2000
- Below 500

**FIGURE 7.8** Physical problems in Peru

## PROBLEMS OF THE COUNTRYSIDE – HUMAN PROBLEMS

Most people migrate to Lima from the Andes mountains where living conditions are harsh. Here the mud-brick houses have no electricity or running water. There is no safe sewage disposal, so rivers are often polluted and spread disease. Secondary schools are only found in towns and have to be paid for. There are few doctors or health centres.

When a farmer dies, it is the custom to divide the land between his sons so, over the years, farms have become smaller and smaller and are now difficult to make a living from. Also, the soil is becoming poorer because, as more trees are cut down for fires, the fertile topsoil is washed away.

**FIGURE 7.9**

| YEAR | CALORIES PER PERSON PER DAY |
|------|------------------------------|
| 1950 | 2270 |
| 1960 | 2260 |
| 1970 | 2251 |
| 1980 | 2166 |
| 1990 | 2037 |
| 1992 | 1882 |

**FIGURE 7.10** *Amount of food eaten in Peru*

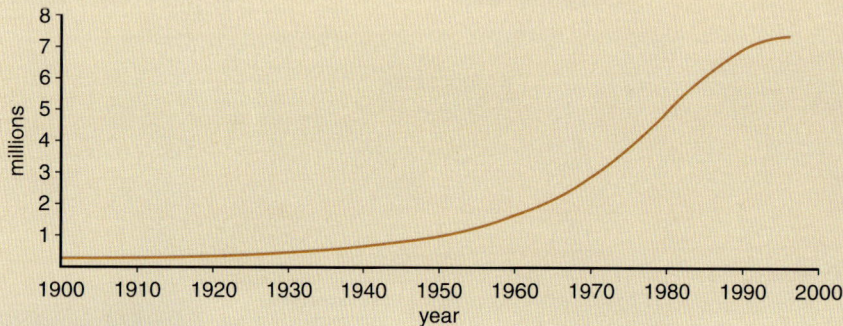

**FIGURE 7.11** *Population of Peru*

## THE ATTRACTIONS OF LIMA

Lima has 70 per cent of all Peru's manufacturing industry and far more shops and entertainments than any other city in Peru. It also has better schools, a university, several well-equipped hospitals and 70 per cent of all the doctors in Peru. Most of the wealthy people live here in expensive houses in the suburbs. For people looking to improve their standard of living, Lima is the place to come.

**FIGURE 7.12**

## EFFECTS OF MIGRATION ON THE COUNTRYSIDE

In some ways people moving away from the countryside brings advantages. Farms should be bigger and more profitable. There should be less pressure on schools and health centres. The people who move away may also send money back to their families. But unfortunately, it is the most educated, most skilled and most dynamic young people who go to the cities. They leave behind the older and less able people who are much less likely to improve conditions in the countryside. So the problems just become worse.

**FIGURE 7.13**

## EFFECTS OF MIGRATION ON LIMA

Although there are many jobs, houses and services in Lima, there is nothing like enough for all the people who now live there. So, most people do not have regular jobs and over one-third live in shanty towns (called 'barriadas'). Some shanties have been improved over the years, but the worse ones have no electricity, no water, no proper sewage disposal and no refuse collection. People's homes are small shacks, made of scrap wood, matting or tin. Most people are hungry, diseases are common and there are no doctors or health centres.

Lima's problems ae made worse by its location in a desert and they are rapidly running out of water. (Their main water supply, the river Rimac, is fully used).

The government spends so much money trying to improve conditions in Lima that it has little left to spend on improving life in the countryside.

**FIGURE 7.14**

**FIGURE 7.15** *Lima*

## 7F A Migration model

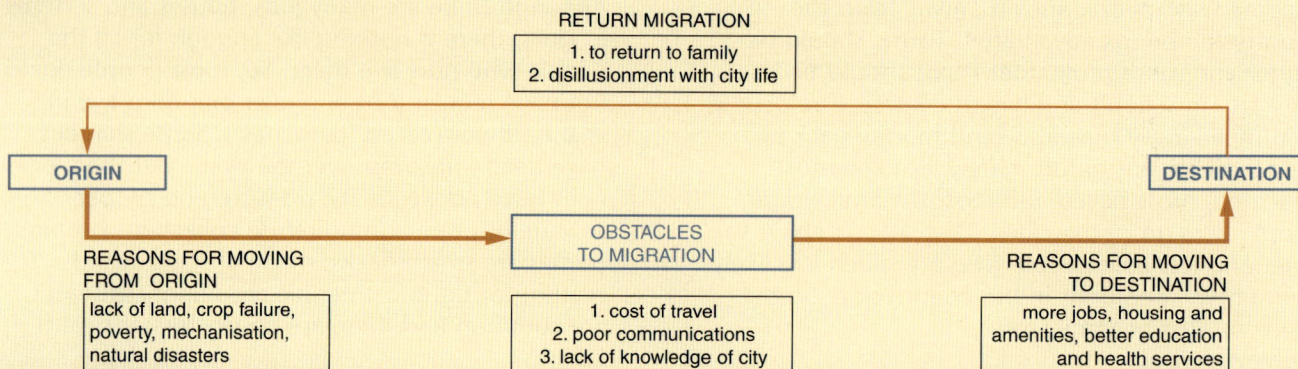

**FIGURE 7.16** A migration model

Figure 7.16 shows a migration model. For any migration movement between two areas, there are push and pull factors which persuade people to leave, but there are also obstacles which deter some people from moving. In addition, some of the people who migrate return to their home area.

## 7G Solving the problems caused by migration

Large-scale rural to urban migration brings major problems to developing countries and all have tried to reduce the worst effects of it. There are several approaches:

1 **Shanty town improvements.** Shanty towns can be improved by making them legal. This makes it easier for people living there to get jobs. If electricity is provided and there are street lights, crime is reduced. If clean water and sewage disposal is provided, fewer people catch diseases.

2 **Sites and services schemes.** Since the authorities cannot afford to build houses for everyone, instead they provide suitable sites with basic services such as electricity and water. The people then build their own houses, often with loans from the authority.

3 **Improve life in the countryside.** If conditions in the countryside can be improved, by providing more health centres and schools and by making farming more profitable, then fewer people will want to leave.

### >> EXTENSION QUESTIONS

*Read the Extension Text.*

**1** Describe some obstacles to migration.

**2** What is meant by return migration and why does it take place?

**3** In what ways can shanty towns be improved?

**4** In what ways might rural improvements affect cities in the developing world?

## Case Study of Peru

**1** **Look at Figure 7.9.**

'More people migrate to Lima from the Andes mountains than the other areas of Peru because the physical problems are greater there.'

Do you agree with the statement above? Give reasons for your answer.

**2** **Look at Figure 7.13.**

Describe the different points of view rural people in Peru might have towards this migration movement.

**3** **Look at Figures 7.9, 7.10 and 7.11**

Each person in Peru eats less food (calories) now than in 1950. Suggest reasons why.

**4** **Look at Figure 7.5 on page 52.**

(a) What message is the cartoon trying to convey?

(b) Do you agree with this message? Give reasons for your answer.

**5** **Look at Figures 7.8 to 7.14.**

Do you think the Peru government should concentrate on improving conditions in Lima or those in the countryside? Give reasons for your answer.

# 8 International Migration

## Core text

### 8A International migration

A country grows in population through births and immigration. A country loses people through deaths and emigration. When people move from one country to another it is called **international migration**. They might move for a few years and then return home. This is called **short-term migration**. If people move permanently it is called **long-term migration**.

People may move because of **push factors** (things they do not like in their own country) or **pull factors** (things that attract them to another country). Usually, people move because they want to (called **voluntary migration**) but, sometimes, they move because they have no choice (called **forced migration**).

### 8B Reasons for voluntary migration

In any country it is common to have immigrants and emigrants. In Britain, people immigrate from other countries in Europe, from India and from Pakistan. People also emigrate from Britain to other countries in Europe, to North America and to Australasia. One of the most common types of migration in recent years has been from developing to developed countries. The main reasons are shown in Figure 8.2 on the following page.

### 8C Effects of migration

The effects of international migration are similar to those mentioned in the last unit.

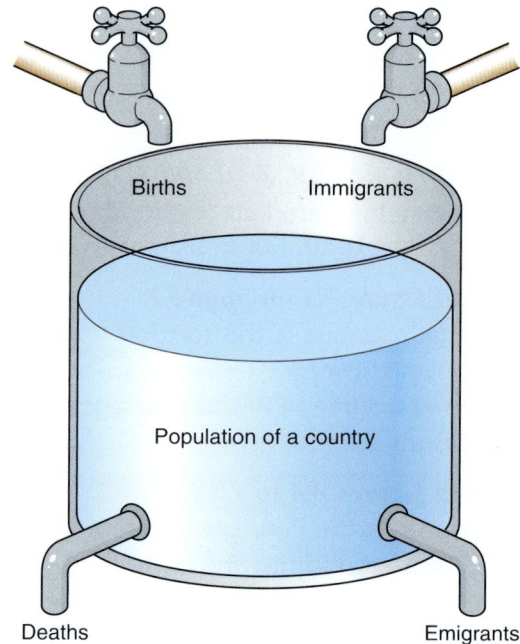

**FIGURE 8.1** *Factors affecting the size of the population of a country*

For the 'losing' country:

1 It loses many 20–35 year olds and their children, so the population rises less quickly.
2 It loses its most go-ahead people.
3 It needs to provide fewer jobs and services and less food.
4 It can cause family break-ups as it is often just the male who migrates.
5 People who move away may send money back home.

For the 'gaining' country:

1 It gains a bigger, often cheaper, labour force (for example Mexicans who work on Californian farms).
2 It needs to provide more services for the extra people (for example hospitals, language classes).
3 The immigrants sometimes set up shanty towns in the cities (for example Moroccans in Marseilles).

**4** It receives new ideas and cultures (for example West Indian music and Asian food in Britain).

**5** Sometimes immigration can lead to racial or religious conflict within the country.

**PUSH FACTORS FROM A DEVELOPING COUNTRY**

**Rural Poverty** low paid farming jobs and few services; soil erosion; unreliable weather

**Urban Poverty** poor housing; disease

**Unemployment** jobs cannot keep pace with the growing population; machines taking over jobs done by people

**PULL FACTORS TO A DEVELOPED COUNTRY**

**Better Education**

**Better paid jobs**

**Better chance of getting a job**

**Better health service**

**More amenities**

**Better housing**

**Government help** government may help people to emigrate

**FIGURE 8.2** *The push and pull factors for voluntary migration*

## 8D Forced migration

People sometimes migrate because they have no choice. This might be because they are being treated unfairly in their own country (called **discrimination**), which might be due to their religion (for example Muslims from Serbia), their race (for example Jews from Germany during World War Two) or their politics (for example Cubans to USA). People are also forced to migrate if there is a war (for example Rwandans into Zaire). When a natural disaster strikes, such as an earthquake, people again have no choice but to leave (for example people fleeing from the volcanic eruption on Montserrat).

People who are forced to move away from their country are called **refugees**. No one knows exactly how many refugees there are in the world, but there are at least 10 million and there may be as many as 50 million.

If only a few refugees enter a new country, it is easier for them to find a better way of life there. If there is mass migration however, the new country will not be able to cope and the refugees will be forced to live in extreme poverty in **refugee camps**. The main characteristics of refugees are listed below:

- 50 per cent are children.

- 80 per cent live in developing countries.

- Most live in extreme poverty, without enough food, shelter, clothing, education or health care.

- They do not belong to their new country but are very unlikely to return to their home country.

- They are extremely unlikely to find work.

- They are not wanted in their new country.

**FIGURE 8.3** *A refugee camp*

1. *Look at 8A.*
   What two things make a country's population grow?

2. What two things make a country's population fall?

3. What is international migration?

4. *Look at Figure 8.2.*
   (a) Describe two problems in developing countries which make people want to emigrate.
   (b) Describe two attractions of a developed country.

5. *Look at 8C.*
   Describe one good effect and one bad effect of a country losing people by emigration.

6. Describe one good effect and one bad effect of a country gaining people by immigration.

7. *Look at 8D.*
   What is discrimination?

8. Give two reasons why people are forced to leave a country.

9. What is a refugee?

10. Describe the conditions in which most refugees live.

## FOUNDATION QUESTIONS

### Case Study of Turks migrating to Germany

1. *Look at Figure 8.6.*
   When did most people migrate from Turkey to Germany?

2. Why did the number of Turkish migrants to Germany decrease after 1973?

3. *Look at Figure 8.7.*
   Suggest why many Turkish farmworkers have emigrated.

4. *Look at Figure 8.8.*
   Why did Germany need more workers after World War Two?

5. *Look at Figure 8.9.*
   Compare the income per person in Turkey and Germany in 1960.

6. *Look at Figure 8.10.*
   When people emigrated from Turkey,
   (a) how did this affect the number of children in Turkey?
   (b) how did it affect unemployment there?

7. *Look at Figures 8.4 and 8.11.*
   What is Figure 8.4 trying to show?

8. *Look at Figure 8.12.*
   What, do you think, was the main reason why some Turks went back home in the 1980s?

**FIGURE 8.4** *Conflicts between Turks and Germans*

## GENERAL QUESTIONS

### Case Study of Turks migrating to Germany

**1** *Look at Figure 8.6.*
Between which years did most Turkish people migrate to Germany?

**2** *Look at Figure 8.7.*
Describe the push factors which made people want to migrate from Turkey.

**3** *Look at Figure 8.9.*
In what ways does Figure 8.9 explain why many Turks migrated to Germany?

**4** *Look at Figure 8.10.*
Do you think the emigration of Turks to Germany has been good or bad for Turkey? Give reasons for your answer.

**5** *Look at Figure 8.11.*
Describe one advantage and one disadvantage to Germany of Turkish immigration.

**6** *Look at Figures 8.4 and 8.11.*
What is the cartoon trying to show?

## Case Study of Turks migrating to Germany

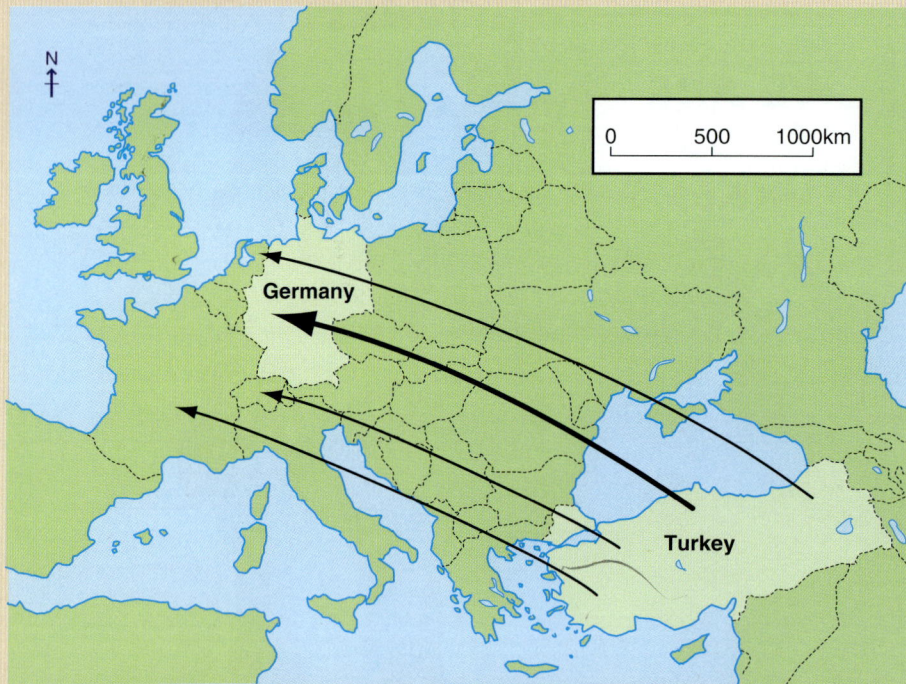

**FIGURE 8.5** *Migration movements of Turkish people in recent years*

### INTRODUCTION

Turkey is a developing country on the Mediterranean Sea and people there have, for many years, migrated to other countries to try and improve their standard of living. Since 1945, most have migrated to the richer countries of Europe and especially to Germany. This migration movement continued until 1973 when foreign workers were banned from entering Germany.

Originally, the Turks were guestworkers (short-term migrants) in Germany, but many have chosen to stay and become long-term migrants.

**FIGURE 8.6**

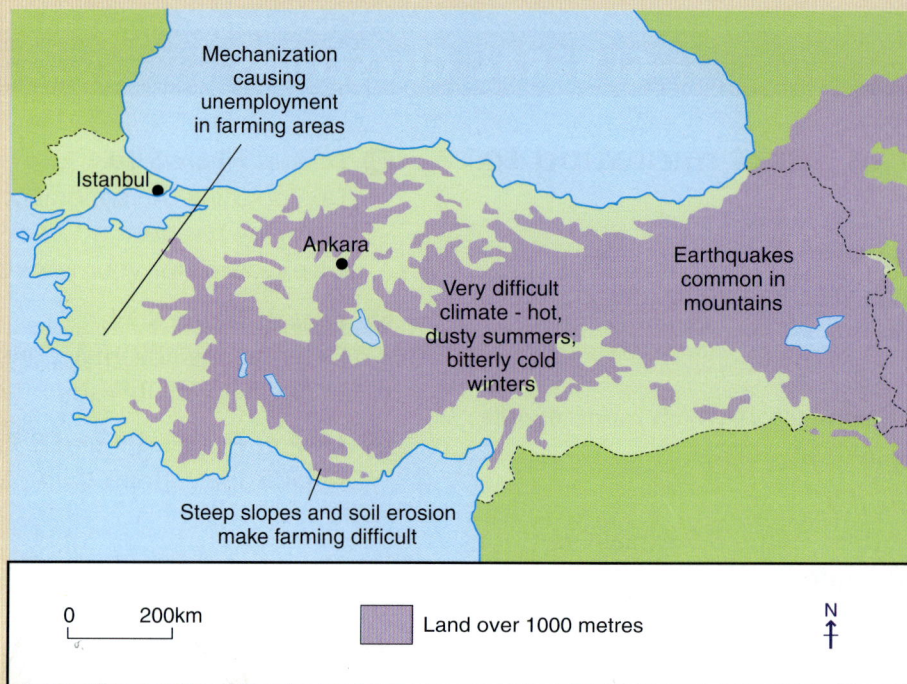

**FIGURE 8.7** *Problems in Turkey*

## THE ATTRACTIONS OF GERMANY

World War Two left Germany with few people of working age. Many parts of the country were also in ruins and much rebuilding work had to be done. Because Germany was short of workers it tried to attract people from Turkey by setting up recruitment offices there. Turkish people were attracted by the opportunity of jobs which, by Turkey's standards, were well paid.

**FIGURE 8.8**

|  | TURKEY (1960) | GERMANY (1960) |
|---|---|---|
| Income per person | £134 per year | £891 per year |
| % working in manufacturing | 5 | 35 |
| Cars per 1000 people | 2 | 69 |
| Infant mortality | 230 | 50 |
| Natural increase per year | 27‰ | 6‰ |

**FIGURE 8.9** *Standards of living in Turkey and Germany in 1960*

## THE EFFECTS ON TURKEY

As people left the countryside, there was less pressure on the land and farms became bigger and more profitable. Because it was mostly young people who left, fewer children were born so the pressure on the land was reduced even more. In the cities, unemployment was reduced as people emigrated. Once they had moved away, many migrants began to send money back to their families (as much as one million pounds per year) and, when they returned, they brought new ideas and skills with them.

However, Turkey lost its most able and educated young adults. In the countryside, especially, there were mostly old people left who were not able to improve the conditions there. Many families were broken up, as it was the male who left to find work. Turks also became worried that so many people were moving away from the east that there were not enough people to defend its eastern border.

**FIGURE 8.10**

### THE EFFECTS ON GERMANY

Germany welcomed Turkish immigrants in the 1950s and 1960s because they were willing to do low paid, dirty, unskilled jobs. Some immigrants also brought valuable new skills and they worked long hours in poor conditions. By the 1980s, 3 per cent of the German workforce was Turkish. The Ford car factory in Cologne, for example, employed 5000 Turks.

The Turks mainly lived in cities, such as Frankfurt and Munich, and in the poor inner city areas of those cities. However, there was not enough housing for them and it was costly for the city authorities to provide more hospitals and schools with special language classes.

During the 1970s and 1980s, there were fewer jobs available as companies were hit by the recession and 20 per cent of Turks became unemployed. Some Germans resented those Turks who still had jobs and racial tension and conflict increased.

**FIGURE 8.11**

### REASONS FOR RETURNING TO TURKEY

Some Turks returned home as soon as they had made some money. Many, however, found the cost of living so high in Germany that they did not make as much money as they had hoped and so stayed in Germany. In 1980, Germany offered Turks grants to return home (up to £2500), but very few have taken up the offer.

**FIGURE 8.12**

**FIGURE 8.13** *Specialist Turkish shop in Germany*

## 8E  Types of migration

| REASON | VOLUNTARY | FORCED |
| --- | --- | --- |
| Economic | To improve standard of living; guestworkers | Slave Trade; Highland Clearances |
| Political | Movement of people to New Towns | Cubans to USA; Vietnamese to Hong Kong |
| Racial/religious | Pilgrimages; Jews to Israel | Movement of Jews in World War Two; segregation of blacks in South Africa |
| Environmental | To enjoy a better climate, for example retired people to the south coast of England | Famine; earthquake; volcanic eruption |
| Personal/social | To be with friends or family; holidaymakers | Settling of nomadic Inuit in permanent villages in northern Canada |

There are many different types of migration. They can be categorised according to the reasons for moving or according to whether they are voluntary or not (as in the table above). They can also be categorised according to the distance moved (for example local, regional, national, international) or frequency of movement (for example daily commuters, seasonal holidaymakers, short-term guestworkers, permanent migrants).

## >> EXTENSION QUESTIONS

*Look at the Extension Text.*

1. Describe the following migrations according to (a) the reasons for moving, (b) voluntary or forced, (c) distance moved, and (d) frequency of movement.

   1. Irish families emigrating to the USA after the potato famine in the 1840s.

   2. African Muslims making a pilgrimage to Mecca in Saudi Arabia.

   3. Greek women joining their husbands who have emigrated to Australia.

   4. The transfer of Civil Service jobs from London to Scotland.

   5. Racehorse jockeys going to a different racecourse every day.

   6. People moving from inner city Edinburgh to the suburbs because of redevelopment.

   7. British engineers with two year contracts in Saudi Arabia.

## Case Study of Turks migrating to Germany

**1** *Look at Figures 8.7, 8.8 and 8.9.*
Many Turks migrated to Germany in the 1950s and 1960s. Do you think the push or the pull factors were more important? Give reasons for your answer.

**2** *Look at Figure 8.10.*
Describe the different points of view Turkish people would have towards the emigration of so many of their people.

**3** *Look at Figures 8.4 and 8.11.*
Describe the conflict shown by Figure 8.4.

**4** *Look at Figures 8.11 and 8.12.*
(a) Suggest why Germany offered the Turks money to return to Turkey in the 1980s.
(b) Suggest why so few Turks took up the offer.

# 9 Skills in Development Studies

## Core text

### 9A Introduction to development studies

As explained in Unit 1, the study of population is very important in Geography. As well as studying where people live, we also need to look at how people live and how people everywhere try to improve their lives. This is called Development Studies and is covered in the remaining units of this book.

For the Standard Grade examination, you need to know and understand the following:

1 Causes and effects of cutting down rainforests, of ocean pollution and of the spread of deserts and how these problems can be solved.
2 Differences in trade patterns between developed and developing countries.
3 Trade problems of developing countries and their solutions.
4 Benefits and problems brought by international alliances, including the European Union.
5 Reasons why international aid is needed.
6 Types of international aid and self-help schemes and their advantages and disadvantages.
7 How to measure the international influence of a country.

You also need to develop the following enquiry skills:

1 How to gather information on development issues by undertaking interviews.
2 How to process this information by drawing bar graphs and pie graphs.

### 9B Preparing an interview

Much of the information on development issues comes from official statistics. It is, however, possible to obtain some firsthand information by interviewing people. With this gathering technique you could obtain detailed information on environmental issues (from a representative of an environmental group), on international trade (from an employee of a large company), on international aid (from a worker in a charity organisation) and on international alliances (from an official of the European Union).

Before undertaking an interview, you should be properly prepared:

● Arrange the interview in advance, by letter or by phone.

● Tell the person the reason for the interview.

● Prepare the questions beforehand and write them down on a recording sheet, leaving space for answers.

● Do not ask unnecessary questions. The questions you ask should allow you to answer your aim. For example, for the interview shown in Figure 9.1 there is no need to ask questions about why rainforests are being cut down or how it can be reduced, because these are not the aims of the interview.

● The questions should allow you to obtain detailed information. For example, for the interview shown in Figure 9.1, if you just ask what the effects of rainforest destruction are, you may either get an incomplete or an extremely long answer. Use your own understanding of the topic to ask more specific questions.

● At the end of the interview, thank the person for his/her help.

Date of interview: 27 February 1999

**Person Interviewed:** Mrs Griffiths, Friends of the Earth

**Aim:** to investigate the effects of rainforest destruction in Brazil

**Questions:**

1 How much of the rainforest in Brazil has been cut down? _____

2 How rapidly is it being cut down? _____

3 What effects has this had on:
   (a) the wildlife _____
   _____

   (b) the soil _____
   _____

   (c) the local Indian people? _____
   _____

4 Is there any evidence that it has affected the climate? _____
   _____

5 Have there been any other problems caused by destroying the rainforest? _____
   _____

6 Has the amount of farmland increased as the rainforest has been cut down? _____
   _____

7 Have people from other parts of Brazil settled here? _____
   _____

8 Has the destruction of the rainforest made it possible to extract minerals? _____
   _____

9 Have some rainforests been destroyed to make way for reservoirs? _____
   _____

10 How much timber does Brazil export? _____
   _____

11 Have there been any other advantages of rainforest destruction? _____
   _____

**FIGURE 9.1**

## 9C  Processing techniques

The information you obtain from an interview, such as that in Figure 9.1, can be presented in the form of a written report. But, some of the information you might research for Development Studies can be shown better in the form of maps, graphs and diagrams. These are called processing techniques and two are now studied in detail.

## 9D  Drawing a bar graph

A bar graph is used to compare the quantity of several different items.

- Draw an *x* (horizontal) and a *y* (vertical) axis and use the *y* axis to show information that is a quantity. In Figure 9.2, for example, the amount of money owed, is shown on the *y* axis.

- Work out a suitable scale for the *y* axis.

- Count up how many 'bars' you must draw along the *x* axis and then decide how wide each 'bar' should be.

- Draw the axes and bars in pencil and go over them in ink once you have checked their accuracy.

- Write a title for the graph.

## 9E  Drawing a divided bar graph

A divided bar graph may be horizontal or vertical and is used to show how one amount is divided up.

- Draw the axes and choose a suitable scale for the bar graph.

- Draw the largest section first at the bottom (or left hand side) for example in Figure 9.3, the population of Germany.

- Then, draw the second largest section next to the first.

- Work out where this section ends by adding its total to the total so far. For example in Figure 9.3, France begins at 22 per cent and so it should end at 22+16 = 38 per cent.

- Complete all the sections.

- Draw the axes and sections in pencil first, in case you make a mistake.

- Label the axes and sections clearly in ink.

- Write a title for the graph.

## 9F Drawing a pie graph

A pie graph is used to show how one amount is shared out.

- Pie graphs are circles divided into slices or segments.

- Each segment shows an amount – the bigger the segment, the bigger the amount.

- Each segment is worked out in degrees and the segments together add up to 360°.

- To draw a pie graph work out or find out the number of degrees in each segment (for example, column 3 in Figure 9.4).

- Draw the first segment in pencil, starting from the top of the circle and going clockwise (see Figure 9.5).

- You might find it easier to start with the smallest segment.

- Work out the size of each segment either by using a protractor or by using a scale around the edge of the circle.

- Then, draw in the second smallest segment next to the first.

- Work out where each segment ends either with a protractor or by adding on its degrees to the total so far.

- Continue drawing segments of increasing size until you have finished.

- Shade in and label each segment.

- Give the pie graph a title.

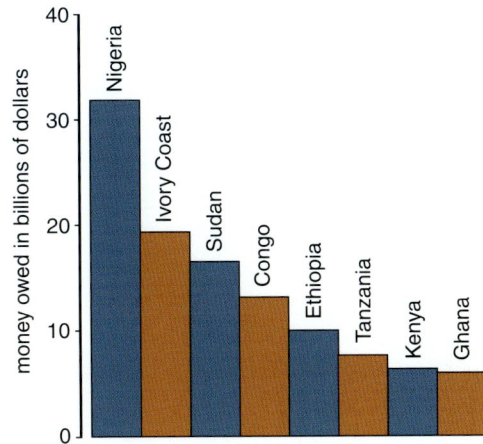

**FIGURE 9.2** *Bar graph showing the debts of eight African countries in 1995*

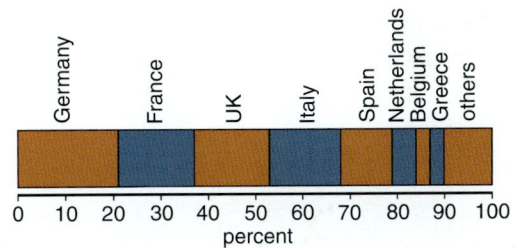

**FIGURE 9.3** *Divided bar graph showing the population of the European Union in 1997*

| NIGERIA'S IMPORTS | COST (£ MILLION) | DEGREES ON PIE GRAPH |
|---|---|---|
| Machinery | 800 | 96 |
| Chemicals | 600 | 72 |
| Foods | 500 | 60 |
| Cars | 300 | 36 |
| Others | 800 | 96 |
| **Total** | 3000 | |

**FIGURE 9.4** *Nigeria's imports*

## 9G Calculating the degrees for a pie graph

- Add up the total number to be represented by all of the segments (for example £3000 million in Figure 9.4).

- For each segment divide its number by the total and then multiply by 360 – this gives you the number of degrees for that segment. For example in Figure 9.4:

$$\text{machinery} = \frac{800}{3000} \times 360 = 96°.$$

- Repeat this calculation for all the segments.
- Check that the total number of degrees adds up to 360.

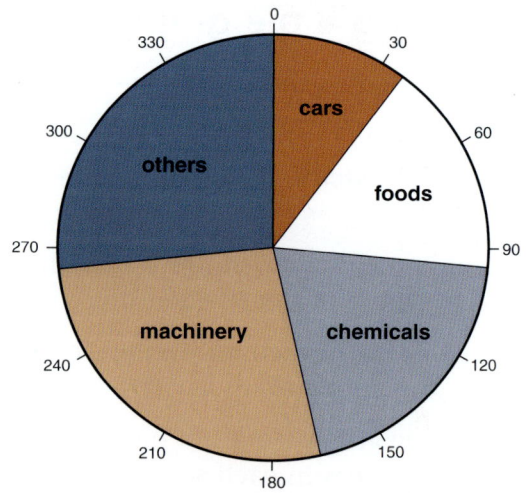

FIGURE 9.5 Pie graph showing Nigeria's imports

# FOUNDATION QUESTIONS

**1** *Look at 9B.*
If you were going to interview a government official on why there is so much pollution in the North Sea, what questions would you ask?

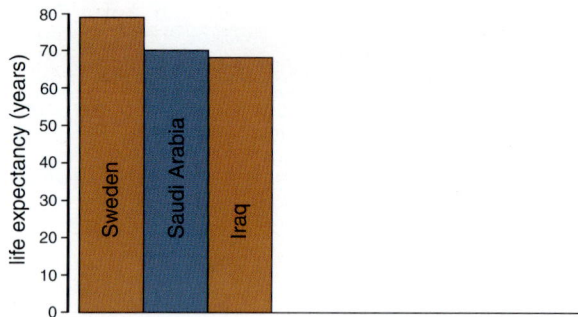

FIGURE 9.6

**2** *Look at 9D.*
(a) Draw the bar graph in Figure 9.6.
(b) Complete the bar graph, using the information in Figure 9.7.

| COUNTRY | AGE |
|---|---|
| Sweden | 79 |
| Saudi Arabia | 70 |
| Iraq | 68 |
| Indonesia | 62 |
| Cameroon | 52 |
| Guinea | 46 |
| Zimbabwe | 41 |
| Zambia | 36 |

FIGURE 9.7 Life Expectancy in Selected Countries

**3** (a) Draw the pie graph as shown in Figure 9.8.
(b) Complete the graph, using the information below.

| TYPE OF IMPORT | PERCENTAGE OF TOTAL IMPORTS | NUMBER OF DEGREES |
|---|---|---|
| oil | 33% | 120 |
| machinery | 29% | 100 |
| foodstuffs | 16% | 60 |
| chemicals | 7% | 25 |
| other manufactured goods | 15% | 55 |

Cuba's Imports (1989)

**4** *Look at Figure 9.9 below.*

Which table of information (A, B or C) would be best shown by (a) a bar graph, and (b) a pie graph?

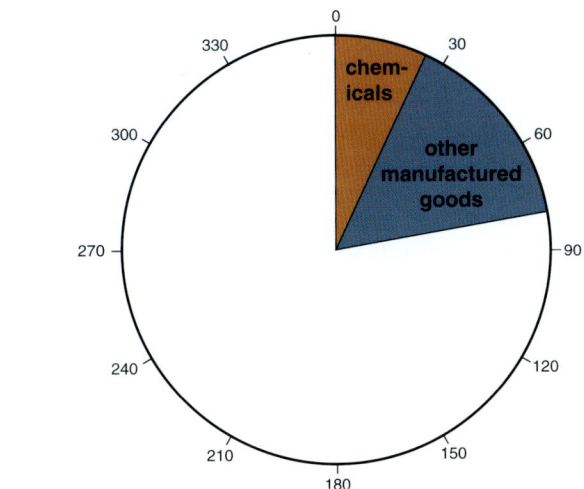

**FIGURE 9.8**

### A
### CAMEROON'S EXPORTS (1989–1994)

| YEAR | VALUE OF EXPORTS (MILLION DOLLARS) |
|---|---|
| 1989 | 130 |
| 1990 | 2080 |
| 1991 | 2900 |
| 1992 | 1930 |
| 1993 | 2120 |
| 1994 | 3000 |

### B
### CAMEROON'S EXPORTS (1991)

| TYPE OF EXPORT | PERCENTAGE OF TOTAL |
|---|---|
| oil | 47% |
| cocoa | 12% |
| wood | 8% |
| machinery | 7% |
| cotton | 4% |
| aluminium goods | 4% |
| coffee | 3% |
| others | 15% |

### C
### EXPORTS OF SELECTED COUNTRIES

| COUNTRY | EXPORT TOTAL (MILLION DOLLARS) |
|---|---|
| Australia | 49 000 |
| Cameroon | 3000 |
| Finland | 30 000 |
| Jamaica | 1000 |
| Pakistan | 7000 |
| Sri Lanka | 3000 |
| Uruguay | 2000 |

**FIGURE 9.9**

# GENERAL QUESTIONS

1. ***Look at 9B.***
If you were going to interview a representative of Oxfam about the effects of a severe drought taking place in part of Africa, what questions would you ask?

2. ***Look at 9D.***
Draw a bar graph to show the information in Figure 9.10.

| NAME OF SHIP (YEAR) | OIL SPILL (TONNES) |
|---|---|
| Amoco Cadiz (1978) | 227 000 |
| Torrey Canyon (1967) | 119 000 |
| Braer (1993) | 85 000 |
| Sea Empress (1996) | 70 000 |
| Exxon Valdez (1989) | 37 000 |

**FIGURE 9.10** *World's Largest Oil Spills*

| TYPE OF EXPORT | PERCENTAGE OF TOTAL EXPORTS | NUMBER OF DEGREES |
|---|---|---|
| tobacco | 69 | 250 |
| tea | 14 | 50 |
| timber | 11 | 40 |
| coffee | 6 | 20 |

**FIGURE 9.11** *Malawi's Exports*

3. ***Look at 9F.***
Draw a pie graph to show the information in Figure 9.11 above.

4. ***Look at Figure 9.9.***
Which table of information (A, B or C) would be best shown by (a) a bar graph and (b) a pie graph? Give a reason for each answer.

**1** *Look at 9B.*
You are researching the causes of the rapid increase in desertification in a West African country and you have arranged to interview an expert from a nearby university. What questions do you ask him?

**2** *Look at 9E.*
Draw a divided bar graph to show the information in Figure 9.12.

| COUNTRY | |
|---|---|
| Australia | 41 |
| Botswana | 17 |
| Congo | 20 |
| Russia | 13 |
| South Africa | 9 |
| other countries | 8 |

**FIGURE 9.12** *World Producers of Diamonds (1995) Diamond production (million carats)*

| REGION | PERCENTAGE OF TOTAL EXPORTS |
|---|---|
| European Union | 38% |
| North American Free Trade Area | 15% |
| Japan | 9% |
| Organisation of Petroleum Exporting Countries | 4% |
| rest of the world | 34% |

**FIGURE 9.13** *Source of World Exports (1995)*

**3** *Look at 9F and 9G.*
Draw a pie graph to show the information in Figure 9.13 above.

**4** *Look at Figure 9.9.*
Which table of information (A, B or C) could be shown by (a) a divided bar graph, and (b) a pie graph? Give a reason for each answer.

# 10 Clearing the Tropical Rainforests

## Core text

### 10A Introduction

All countries try to improve the standard of living of their people. This is called **development**. For countries beside the equator, one way to develop is to clear their large areas of rainforest (called **deforestation**). But the rainforest is a fragile environment, clearing it brings problems as well as benefits.

### 10B Rainforest environments

See Figure 10.2.

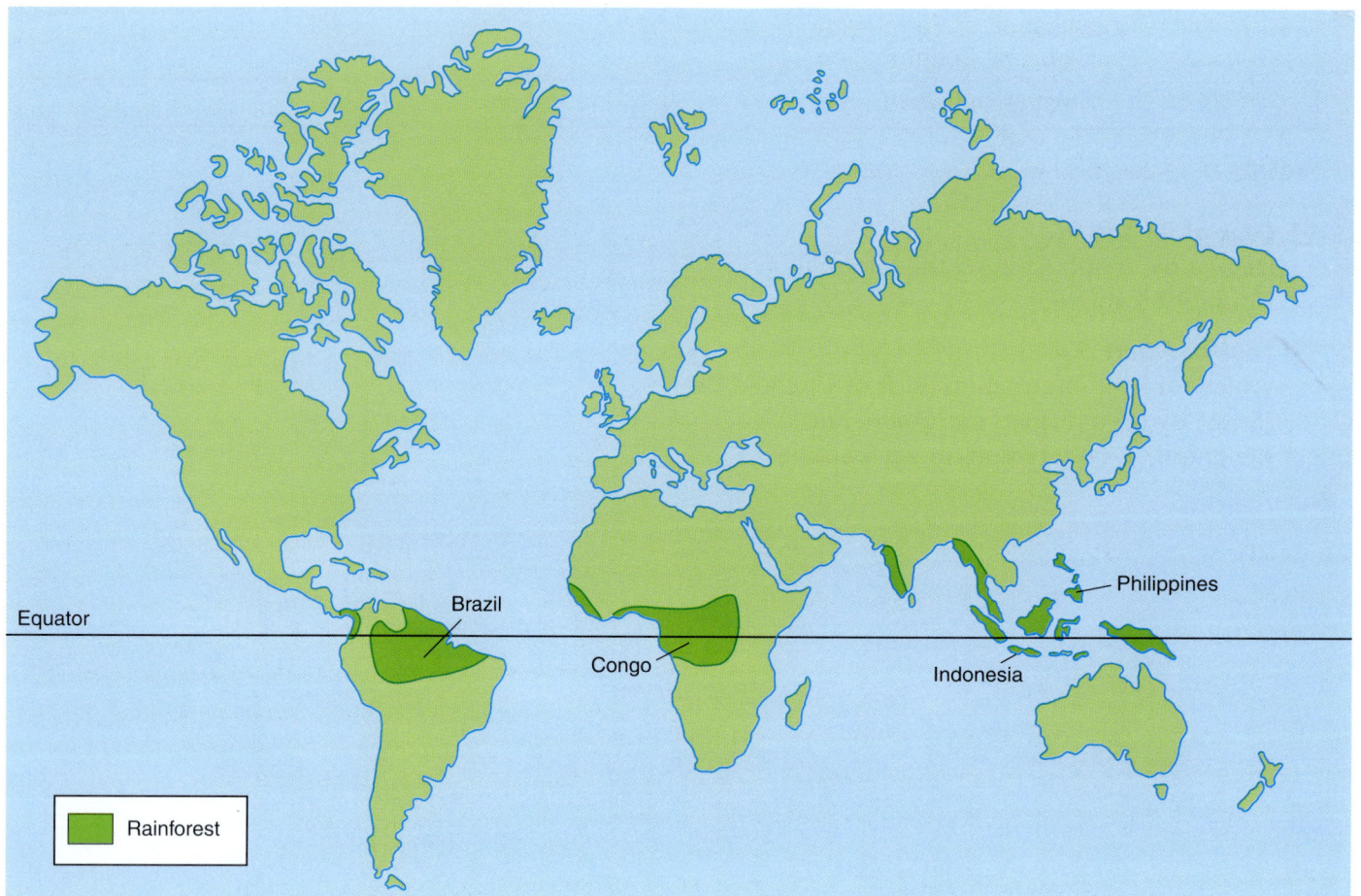

**FIGURE 10.1** *Distribution of tropical rainforests*

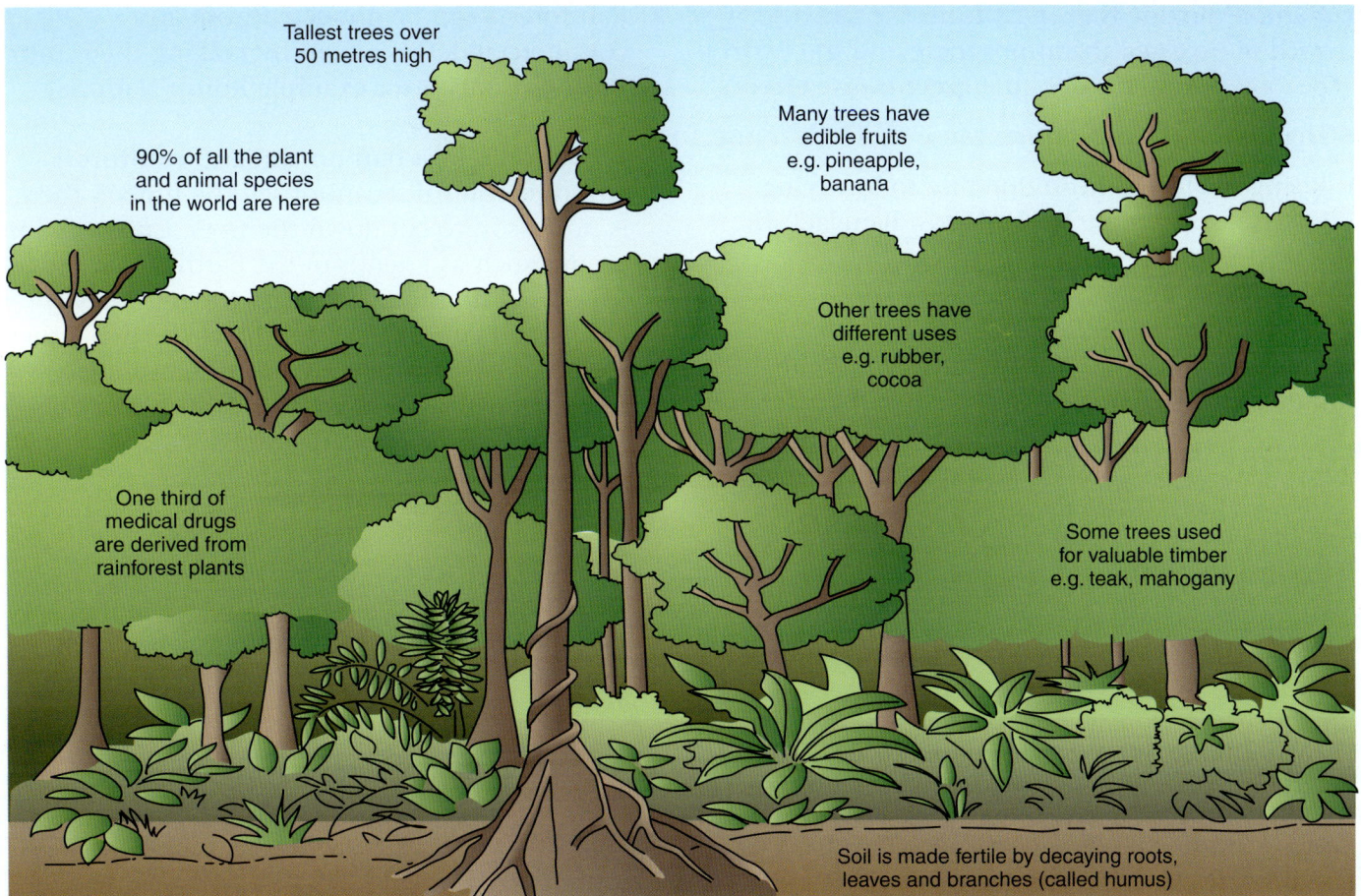

**FIGURE 10.2** *Characteristics of the tropical rainforest*

Labels in figure:
- Tallest trees over 50 metres high
- 90% of all the plant and animal species in the world are here
- Many trees have edible fruits e.g. pineapple, banana
- Other trees have different uses e.g. rubber, cocoa
- One third of medical drugs are derived from rainforest plants
- Some trees used for valuable timber e.g. teak, mahogany
- Soil is made fertile by decaying roots, leaves and branches (called humus)

## 10C Why rainforests are being cleared

Most of the world's rainforests lie within developing countries, and these countries need to find ways of increasing their wealth. From their point of view, clearing rainforests brings the following benefits:

- It makes extra farmland to grow more food.
- It makes it possible for people to settle here, away from overcrowded areas of the country.
- Any minerals can be mined more easily.
- Reservoirs and hydroelectric power stations can be built.
- The felled trees can be sold as timber.
- More roads and railways can be built.

## 10D Effects of deforestation

1 **Effects on the soil:**

- With no leaves or roots, the soil becomes very poor.
- With no tree roots, the topsoil is easily washed away by the heavy rains.
- The soil becomes too thin and poor for crops to grow well.

2 **Effects on the wildlife:**

- Plants used to make valuable medicines die out.
- The homes of thousands of species of birds and animals disappear.
- The wildlife dies or moves away and some species may become extinct.

3 **Effects on the climate:**

- The amount of carbon dioxide in the air increases because there are fewer trees to take in this gas.

- Carbon dioxide traps heat from the Earth so, with more $CO_2$, the atmosphere and the Earth become warmer (called the **greenhouse effect**).

- Higher world temperatures cause sea levels to rise.

- Rising sea levels mean flooding to low-lying coasts, for example Netherlands, Bangladesh.

**4  Effects on the local people:**

- People who make their living from the rainforest (for example by hunting animals, tapping rubber trees) lose their livelihood.

- People who live in the rainforest lose their whole way of life.

- Some people are moved into reservations, against their wishes.

- People come into contact with outsiders who spread diseases to which they have no immunity.

## 10E  How to reduce deforestation

In the 1980s, ten per cent of all the tropical rainforest in the world was destroyed. In the 1990s the rate of destruction has become even faster. These are some of the ways of reducing deforestation or the harmful effects of it:

1  In each area of forest, work out how many trees can be cut down without damaging the environment. These trees can then be cut down as long as the same number of trees are planted. This is called **sustainable forestry**.

2  Rainforests can be protected from any commercial development by making them into **national parks**, for example Korup National Park in Cameroon.

3  There could be a **ban on trade in rainforest wood**. If countries cannot sell the timber, there is no reason to cut down the trees. Conservation groups such as Friends of the Earth already urge people not to buy wood from the rainforests.

4  Developed countries could forget the debts they are owed by developing countries, as long as these countries preserve their rainforests. This scheme is called **debt for nature**.

**Mahogany is MURDER Don't buy it**

**FIGURE 10.3** *How a Conservation group's poster might look*

## CORE QUESTIONS

1  *Look at 10A.*
What is meant by 'deforestation'?

2  *Look at 10B.*
Name two fruits that grow in tropical rainforests.

3  What percentage of the world's plant and animal species are found in the rainforests?

4  What fraction of medical drugs come from rainforest plants?

5  *Look at 10C.*
Give three reasons why people clear rainforests.

6  *Look at 10D.*
When the rainforests are cleared,

(a) why does the soil become poorer?
(b) why does the amount of carbon dioxide rise?
(c) how does carbon dioxide make the Earth warmer?

7  Describe two ways in which cutting down rainforests affects the local people.

8  *Look at 10E.*
Name one way in which the world's rainforests can be protected from any development.

9  In a sustainable forestry development, how do they decide how many trees can be cut down?

10  What is the 'debt for nature' scheme?

## Case Study of the Amazon rainforest

**1** *Look at Figure 10.6.*
How quickly is the Amazon rainforest being cut down?

**2** *Look at Figure 10.8.*
Which country of the Amazon rainforest (a) owes most money? (b) is growing most quickly?

**3** *Look at Figure 10.7.*
Why do the countries of the Amazon rainforest need to develop their rainforests?

**4** *Look at Figure 10.12.*
Name two ways in which developments in the Amazon rainforest provide these countries with money.

**5** From where have most people come who now live in the Amazon rainforest?

**6** *Look at Figure 10.13*
After trees in the Amazon have been cut down,
(a) why is there less wildlife?
(b) why are the soils poor?
(c) why are there fewer Indian tribes?

**7** *Look at Figure 10.11.*
Do you think the Carajas Project in Brazil is a good idea? Give reasons for your answer.

**8** *Look at Figure 10.4.*
Do you agree with the statement in the cartoon? Give reasons for your answer.

**9** *Look at Figure 10.14.*
Do you think it is a good idea to set up more National Parks in Brazil? Give reasons for your answer.

**FIGURE 10.4**

## Case Study of the Amazon rainforest

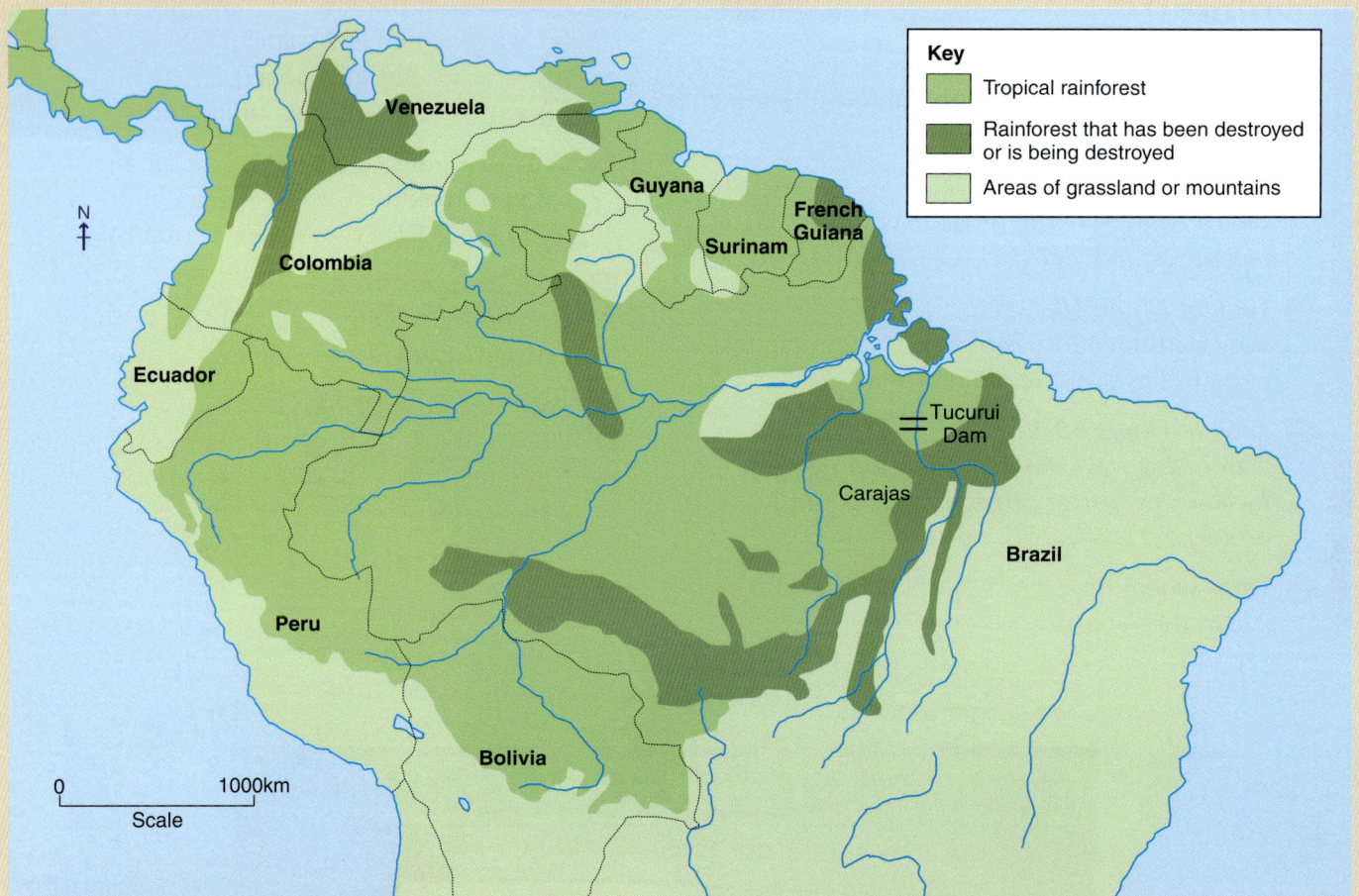

### Key
- Tropical rainforest
- Rainforest that has been destroyed or is being destroyed
- Areas of grassland or mountains

**FIGURE 10.5** *Developments in the Amazon rainforest*

### INTRODUCTION

The Amazon rainforest is in South America and 75 per cent of it is in the country of Brazil. It covers more than five million square kilometres and is by far the largest area of rainforest in the world, but it is being cut down rapidly. In the last 40 years, an area of forest twice the size of Britain has been destroyed.

**FIGURE 10.6**

### REASONS FOR DEFORESTATION

In the countries of the Amazon rainforest the population is growing very rapidly, there is not enough land for farmers and not enough jobs in the cities. They are poor countries and all are in debt to developed countries. By exploiting the resources in the rainforest, these countries can make money and pay off their debts.

**FIGURE 10.7**

| COUNTRY | POPULATION (MILLIONS) | AVERAGE ANNUAL INCOME (£) | EXTRA PEOPLE EACH YEAR | INTERNATIONAL DEBTS (MILLIONS OF DOLLARS) |
|---|---|---|---|---|
| Bolivia | 8.0 | 500 | 46 000 | 5266 |
| Brazil | 160.0 | 2300 | 1 276 000 | 159 130 |
| Colombia | 36.0 | 1200 | 430 000 | 20 760 |
| Ecuador | 12.0 | 900 | 189 000 | 13 957 |
| French Guiana | 0.2 | 4100 | 11 000 | – |
| Guyana | 0.8 | 400 | 3000 | 2105 |
| Peru | 26.0 | 1400 | 333 000 | 30 831 |
| Surinam | 0.4 | 600 | 6000 | – |
| Venezuela | 22.0 | 1900 | 410 000 | 35 842 |

**FIGURE 10.8** *Standard of living in the countries of Amazonia*

## DEVELOPMENTS IN THE BRAZILIAN RAINFOREST SINCE 1960

- Thousands of kilometres of roads have been built.

- People have been given free or very cheap land to farm.

- Cattle ranching is the most common type of farming, requiring large areas of land.

- Rivers are dammed to make reservoirs and HEP stations.

- Minerals are extracted, for example tin, bauxite, iron ore (see Figure 10.5).

- Population has increased ten fold.

- 85 per cent of original rainforest is still left.

- Rainforest covers 40 per cent of Brazil, but still contributes only three per cent to its national income.

**FIGURE 10.9**

## THE CARAJAS PROJECT, BRAZIL

Starting in 1985, this project aims to develop the eastern part of Brazil's rainforest, by:

- Mining the huge deposits of iron found there, together with other minerals.

- Making the iron ore into pig iron at 15 iron-smelting plants, which use charcoal for fuel – this requires 1500 sq km of forest to be felled each year.

- Building a large HEP station at Tucurui.

- Attracting industries because of the available iron, HEP and government subsidies.

The lands of two Indian tribes will be lost by this project and Brazil has had to borrow 600 billion dollars from the European Union to pay for it.

**FIGURE 10.11**

# RESOURCES

**FIGURE 10.10** *Carajas iron ore mine, Amazonia, Brazil*

## EFFECTS OF DEFORESTATION IN THE AMAZON BASIN

1 It has enabled many people to move here from overcrowded areas and from areas suffering drought and famine.

2 There are millions of landless people in this region who are now able to own and work land.

3 It has made it possible to mine valuable minerals which are then exported.

4 Timber can also be exported, providing these countries with much needed money.

5 Hydroelectric power stations have attracted industries which provide many jobs. They also save the countries importing other fuels such as coal and oil.

6 Deforestation has also caused many problems and these are shown in Figure 10.13.

**FIGURE 10.12**

## SOLUTIONS

1 In countries such as Colombia and Brazil, national parks have been set up. Within these areas, developments such as mining, cattle ranching and road building are not allowed, although tourism is encouraged.

2 In Brazil, landowners must keep half of their land as forest. This has not been successful as the owners later sell the land which is still forested.

3 Brazil has stopped giving tax benefits to people who farm the rainforests.

4 In countries such as Ecuador, international charities have bought small areas of rainforest and will protect them from development.

5 In Colombia, the government has given 60 000 sq km of land back to the local Indian people. It had been wrongly taken from them by a mining company.

6 Developed countries are less likely now to fund projects which are very harmful to the rainforest environment although they are still involved in many schemes.

**FIGURE 10.14**

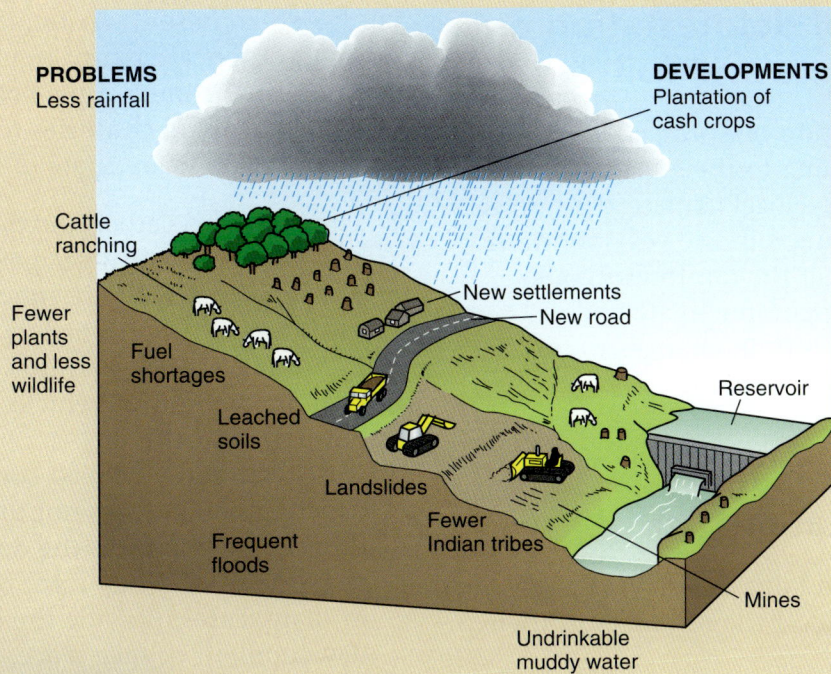

**FIGURE 10.13** *Problems caused by deforestation*

Labels in figure:

PROBLEMS
Less rainfall

DEVELOPMENTS
Plantation of cash crops

Cattle ranching

Fewer plants and less wildlife

Fuel shortages

Leached soils

Landslides

Frequent floods

Fewer Indian tribes

New settlements
New road

Reservoir

Mines

Undrinkable muddy water

## GENERAL QUESTIONS

## Case Study of the Amazon rainforest

**1** *Look at Figure 10.6.*
Describe the size of the Amazon rainforest and the rate at which it is being cleared.

**2** *Look at Figure 10.8.*
Compare the international debts of the countries of the Amazon rainforest.

**3** *Look at Figure 10.7.*
In what ways do these debts explain why the countries need to clear the Amazon rainforest?

**4** *Look at Figure 10.11.*
Describe one advantage and one disadvantage to Brazil of the Carajas Project.

**5** *Look at Figure 10.13.*
Suggest why clearing the Amazon rainforest has led to:
(a) landslides?
(b) fewer Indian tribes?

**6** *Look at Figure 10.4.*
Do you agree with the statement in the cartoon? Give reasons for your answer.

**7** *Look at Figure 10.14.*
Do you think Brazil is likely to set up more National Parks in its rainforest? Give reasons for your answer.

## 10F Effects of deforestation on the climate and soil

Large-scale deforestation not only changes the world climate by adding to the greenhouse effect, it may also change the local climate. This, in turn, will lead to changes in the vegetation. In time, it is thought that the rainforests could become deserts. Changes in climate and vegetation will also result in changes to the soil, as shown in Figure 10.15 below.

## 10G Arguments for deforestation

We have already found out that developing countries need to exploit the resources in their rainforests to raise their standard of living. Although there are many arguments against deforestation, some people believe it is unfair to expect developing countries to stop clearing their rainforests. These are some of their arguments:

1  The increased carbon dioxide in the atmosphere is not principally due to deforestation. It has been caused mostly by fossil fuels being burned in power stations and in motor vehicles, and most of the power stations and motor vehicles are in the developed world.

2  It is unfair of developed countries to tell developing countries not to exploit the rainforests when it is the developed countries themselves who buy most of the products and, in many cases, provide the funds that allow development to take place.

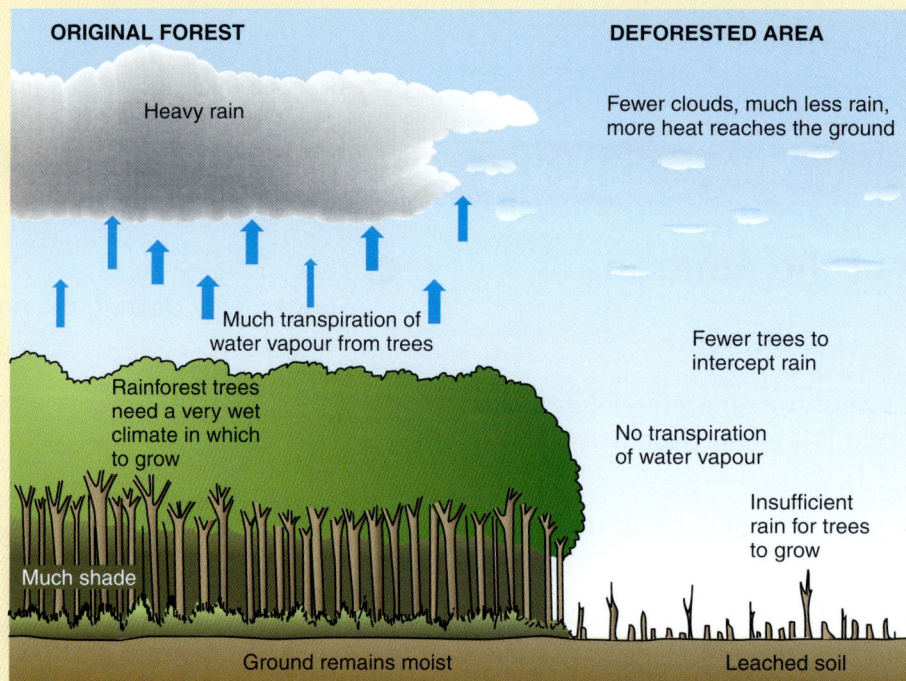

**ORIGINAL FOREST**

Heavy rain

Much transpiration of water vapour from trees

Rainforest trees need a very wet climate in which to grow

Much shade

Ground remains moist

**DEFORESTED AREA**

Fewer clouds, much less rain, more heat reaches the ground

Fewer trees to intercept rain

No transpiration of water vapour

Insufficient rain for trees to grow

Leached soil

**FIGURE 10.15** *Effects of deforestation*

*Look at the Extension Text.*

**1** Explain why rainfall is reduced when large areas of rainforest are cleared.

**2** In what ways does the reduced rainfall and cloud cover affect the vegetation and temperature?

**3** How is soil quality affected by rainforest clearance?

**4** What is the main reason for an increase in carbon dioxide in the atmosphere?

**5** In what ways can developed countries be said to be responsible for rainforest clearance?

## CREDIT QUESTIONS

### Case Study of the Amazon rainforest

**1** *Look at Figure 10.8.*
In what ways does the information in Figure 10.8 explain why the Amazon rainforest is being destroyed so quickly?

**2** *Look at Figure 10.9.*
Do you think recent developments in its rainforest have helped Brazil very much? Give reasons for your answer.

**3** *Look at Figure 10.11.*
Describe the advantages and disadvantages of the Carajas Project.

**4** *Look at Figure 10.13.*
Suggest why cutting down the Amazon rainforest has led to:
(a) unreliable river and lake water.
(b) reduced rainfall.

**5** *Look at Figures 10.8 to 10.13.*
Do you think Brazil would be keen to start a 'debt for nature' scheme? Give reasons for your answer.

# Core groupwork

### The rainforest game

The rainforest shown in Figure 10.16 belongs to one country. The country is going to develop its rainforest. It has decided that:

50 squares should be left as natural forest
25 squares made into farmland
10 squares for new settlements
10 squares for new industries (mines and factories)
 5 squares for new reservoirs/power stations.

TASK 1   Working in groups, discuss which squares are best for farming, settlement, industry and reservoirs.

TASK 2   Draw a rough copy of the grid above on paper.

TASK 3   Each group takes it in turn to call out a square and suggest its land use (farmland, forest, settlement, industry or reservoir). If correct the group scores 10 points.

The teacher will explain which are the correct suggestions.

TASK 4   When a correct land use has been given, put the answer on your copy and make a note of the reason.

TASK 5   When all the squares have been correctly named, the group with the highest score is the winner.

TASK 6   Write a report explaining why the forest has been developed in that way.

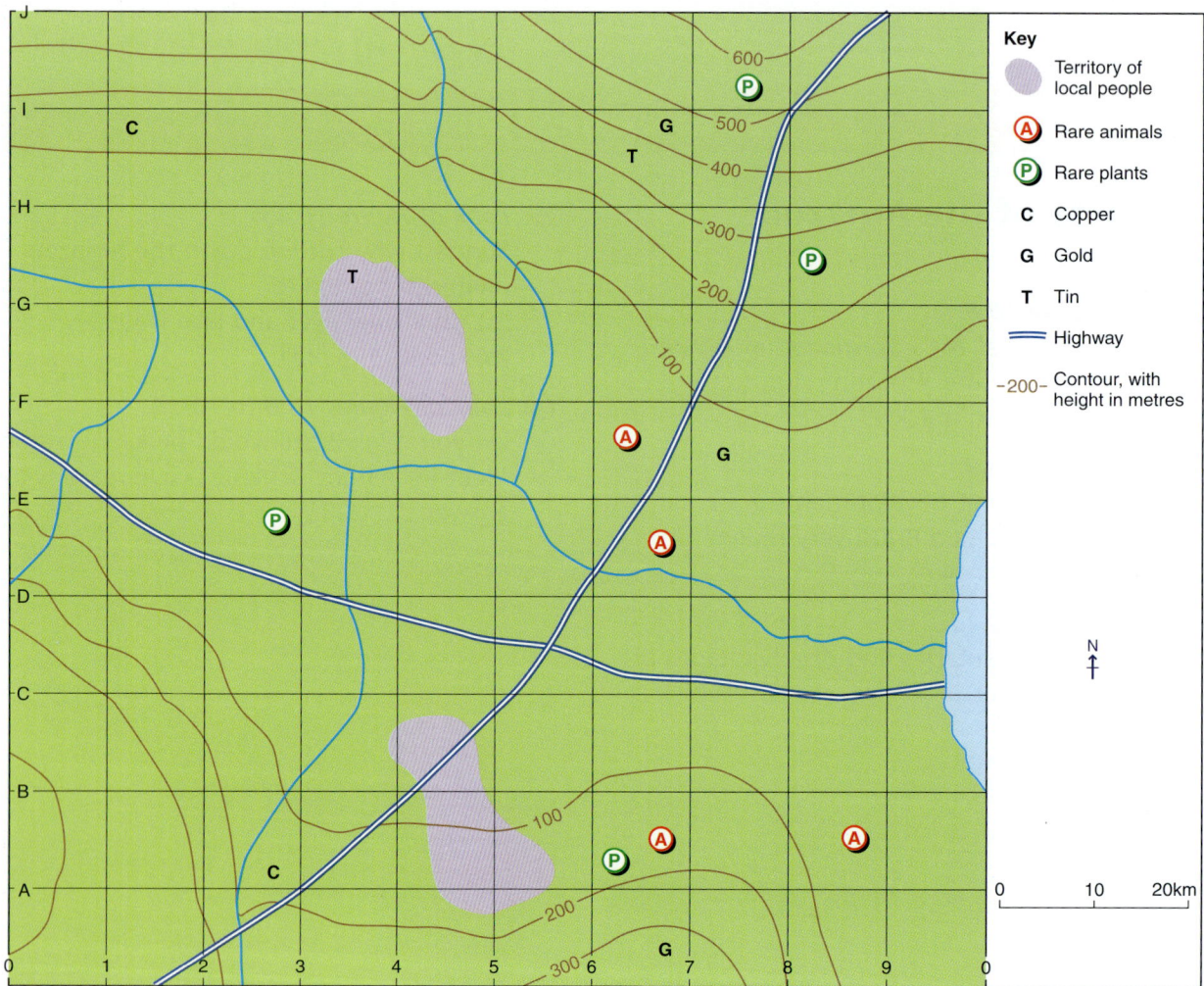

**FIGURE 10.16**

# 11 Using the Seas and Oceans

## Core text

### 11A Uses of the seas and oceans

Just like the tropical rainforests our seas and oceans are being used more and more. They are used for mining, fishing, transport, recreation and dumping waste. As well as bringing benefits, using the seas and oceans for these purposes also brings problems.

### 11B Sea and ocean pollution

One of the main problems caused by using the seas and oceans so much is **pollution**. Pollution comes from four sources: farming, industry, towns and transport. Oceans and seas may be polluted directly or they may become polluted if the rivers that flow into them are polluted themselves. The main pollution blackspots around the world are shown in Figure 11.1. Different types of pollution give rise to a variety of problems, which are shown in Figure 11.2.

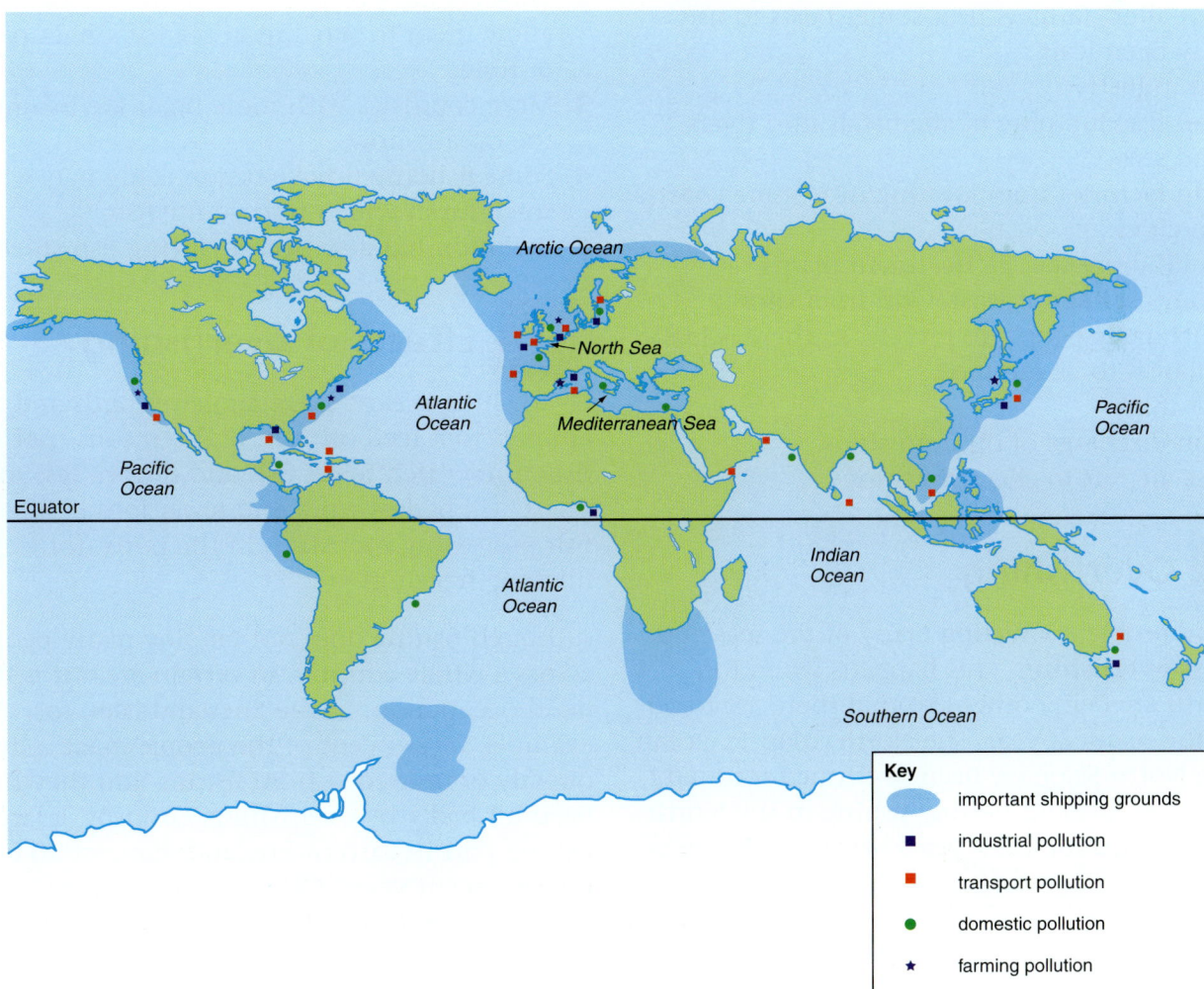

**Key**

- important shipping grounds
- ■ industrial pollution
- ■ transport pollution
- ● domestic pollution
- ★ farming pollution

**FIGURE 11.1** *Important fishing grounds and pollution blackspots*

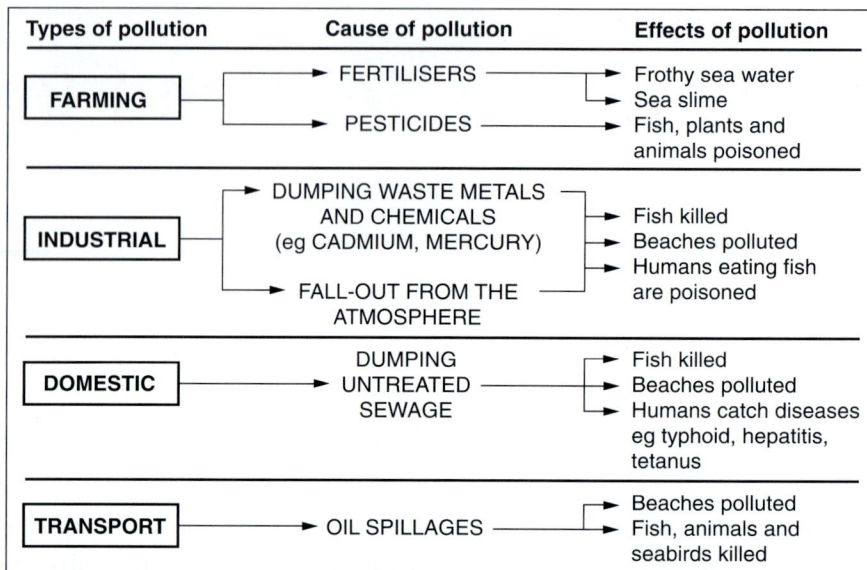

| Types of pollution | Cause of pollution | Effects of pollution |
|---|---|---|
| FARMING | FERTILISERS | Frothy sea water<br>Sea slime |
| | PESTICIDES | Fish, plants and animals poisoned |
| INDUSTRIAL | DUMPING WASTE METALS AND CHEMICALS (eg CADMIUM, MERCURY) | Fish killed<br>Beaches polluted<br>Humans eating fish are poisoned |
| | FALL-OUT FROM THE ATMOSPHERE | |
| DOMESTIC | DUMPING UNTREATED SEWAGE | Fish killed<br>Beaches polluted<br>Humans catch diseases eg typhoid, hepatitis, tetanus |
| TRANSPORT | OIL SPILLAGES | Beaches polluted<br>Fish, animals and seabirds killed |

**FIGURE 11.2** *Causes and effects of pollution*

## 11C How to prevent sea and ocean pollution

1 From farms:
- ban dangerous farm chemicals
- not allow farmers near some coasts to use any chemicals.
2 From industry:
- ban the dumping of chemicals into rivers and seas
- ban factories from locating near some coasts.
3 From cities:
- build more sewage treatment plants
- dump sewage at least 4 km from shore
- encourage people to recycle, so there is less rubbish to be dumped.
4 From ships:
- ban discharge of oil from ships
- use aircraft to check on ships.

## 11D Overfishing

A second problem affecting many of the seas and oceans is **overfishing**. This happens when too many fish are being caught so that there are fewer fish in the sea next year. The North Atlantic Ocean and the North Sea have been overfished for many years and, as a result, herring fishing in the North Sea was banned for a few years because fish stocks were so low.

There are many reasons why overfishing is occurring in the seas and oceans:

1 There are now better methods of finding fish, using echo sounders, satellites and sonar.
2 Trawlers have become larger and factory ships allow them to stay out at sea for longer periods of time.
3 More countries with more boats are fishing all the oceans now.
4 Some fishermen have begun using nets with small mesh which trap young fish.
5 Pollution has also been reducing fish stocks.

## 11E Effects of overfishing

Overfishing means there are fewer birds and animals that feed off fish. It also means that fishermen catch fewer fish and so earn less money. There will be less work and fewer jobs in all the fish processing industries in the ports (for example filleting, freezing).

Although fishing does not employ many people across a whole country, in certain areas it is the main occupation. In the Shetland Isles, for example, 20 per cent of the people make a living directly or indirectly from fishing and they have been hit badly by overfishing. Grimsby, a large fishing port in eastern England, has lost 50 000 jobs in recent years. Other countries have been equally affected. The Danish fishing fleet has been cut by 20 per cent in the last 10 years.

## 11F Solutions to overfishing

1 Many countries now do not allow any other countries to fish within 320 km of their coasts. This should mean fewer boats, and therefore fewer fish caught.

2 Checks can be made to ensure that fishing nets have a wide mesh. Then young fish will survive to breed and increase future stocks.

3 Put a limit or **quota** on how many fish each fishing vessel can catch.

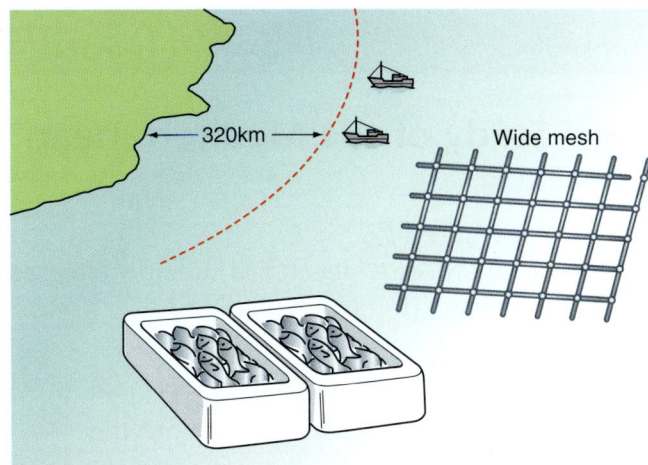

### CORE QUESTIONS

1 *Look at 11A.*
Name four ways in which seas and oceans are used.

2 *Look at Figure 11.1.*
Which types of pollution affect the Atlantic Ocean?

3 Which seas and oceans suffer most from farming pollution?

4 *Look at 11B.*
How can polluted rivers cause seas and oceans to be polluted?

5 *Look at Figure 11.2.*
In what ways does farming cause pollution?

6 What are the effects of industrial pollution?

7 *Look at 11C.*
Describe two ways in which pollution from cities can be reduced.

8 Describe two ways in which pollution from ships can be prevented.

9 *Look at 11D.*
What is overfishing?

10 Give two reasons why overfishing is taking place.

11 *Look at 11F.*
(a) What is a fish quota?
(b) How does it help to reduce overfishing?

### FOUNDATION QUESTIONS

## Case Study of the Mediterranean Sea

1 *Look at Figure 11.4.*
What types of pollution affect the Mediterranean Sea?

2 Name four cities on the Mediterranean Sea.

3 *Look at Figure 11.6.*
(a) How many people live near the Mediterranean Sea?
(b) How many people visit each year?
(c) How do they cause pollution?

4 Which rivers take farm chemicals into the Mediterranean?

5 *Look at Figure 11.8.*
Why, do you think, some wildlife in the Mediterranean is dying out?

6 Why are local people angry about oil spills?

7 *Look at Figure 11.3.*
Do you agree with the farmer's statement in the cartoon? Give reasons for your answer.

8 *Look at Figure 11.9.*
What is being done to reduce the amount of (a) sewage and (b) pollution from factories in the Mediterranean?

9 *Look at Figure 11.10.*
Describe two ways of dealing with oil spills in the Mediterranean.

10 *Look at Figure 11.12.*
Describe one way in which overfishing in the Mediterranean can be reduced.

## Case Study of the Mediterranean Sea

**1** *Look at Figure 11.6.*
Why is the Mediterranean Sea so badly affected by (a) sewage (b) oil pollution?

**2** *Look at Figure 11.7.*
Why does it take so long for any pollution in the Mediterranean to be washed into the Atlantic Ocean?

**3** *Look at Figure 11.8.*
Many tourist resorts on the Mediterranean are affected by pollution from sewage and oil pollution.

Which, do you think, is the more serious problem for them? Give reasons for your answer.

**4** *Look at Figure 11.3.*
Do you agree with the farmer's statement in the cartoon? Give reasons for your answer.

**5** *Look at Figure 11.10.*
Which of the three methods of dealing with oil spills is the best? Give reasons for your answer.

**6** *Look at Figure 11.12.*
Describe two ways in which the Common Fisheries Policy reduces overfishing.

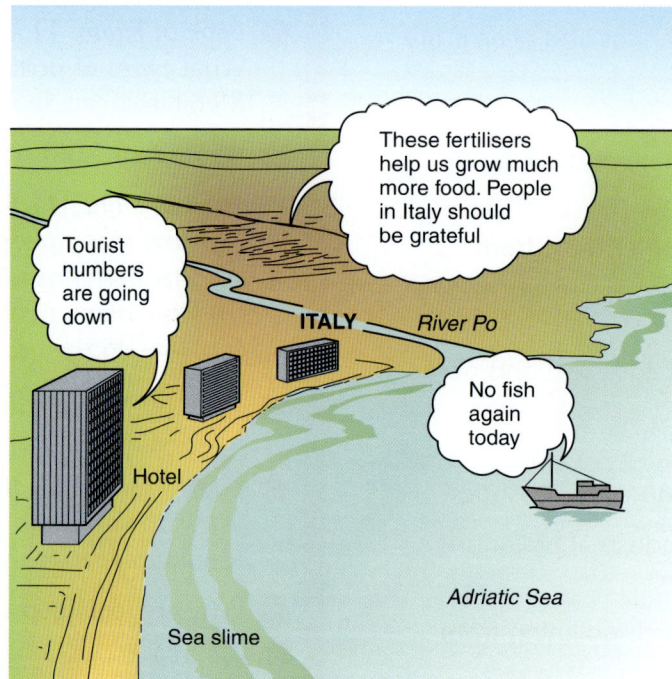

**FIGURE 11.3**

## Case Study of the Mediterranean Sea

### INTRODUCTION

There are 21 countries that border the Mediterranean Sea and all of them use it to dump their waste, for fishing and for tourism. But the Mediterranean is a fragile environment and, by using it so much, many problems have developed. The biggest of these by far is pollution.

**FIGURE 11.5**

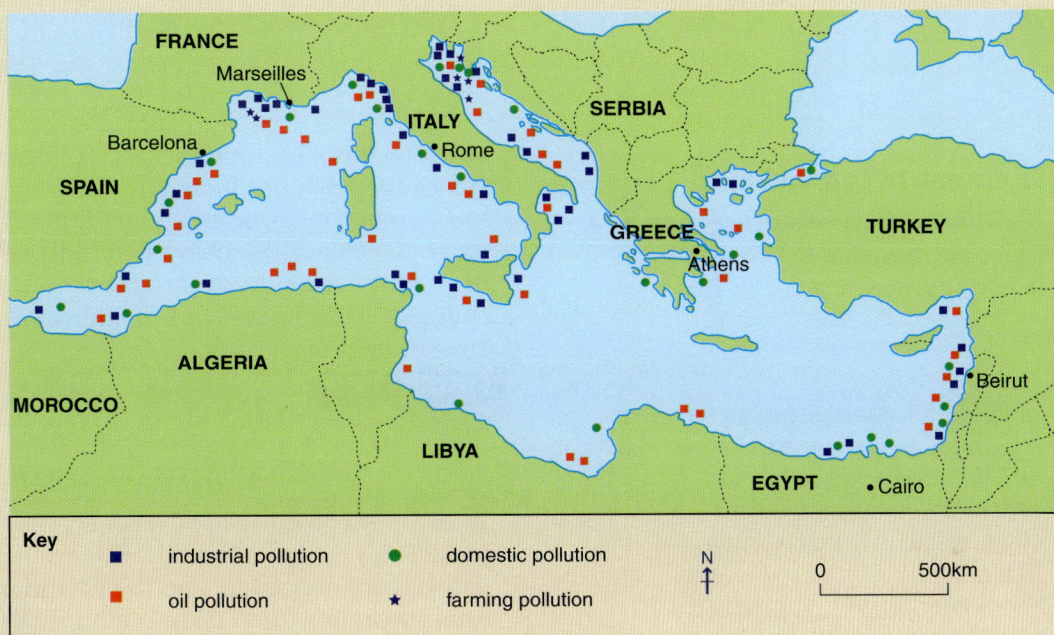

**FIGURE 11.4** Pollution in the Mediterranean Sea

### HUMAN CAUSES OF POLLUTION

- 133 million people live along the Mediterranean coast and they are joined by 120 million tourists each year. Their sewage is dumped in the sea and most of it has not been treated.

- There are 120 cities bordering the Mediterranean. All have factories which dump waste directly into the sea or into rivers which flow into the sea, especially zinc, lead, cadmium and mercury.

- Land near the coast is farmed using many chemicals in order to provide food for the large population nearby. These chemicals get washed into the sea from 70 major rivers such as the Rhone, Po and Nile.

- Oil tankers use the Mediterranean as a quick route between the Middle East and Europe. At any one time there are 300 tankers on the sea and together they deposit 650 000 tonnes of oil each year.

**FIGURE 11.6**

# RESOURCES

## PHYSICAL CAUSES OF POLLUTION

There are two physical characteristics of the Mediterranean which make the problem worse.

- It has weak tides so that any pollution is not easily washed away from the shore.

- There is only a narrow outlet of water into the Atlantic Ocean, at the Straits of Gibraltar, which means water and any pollution takes up to 80 years to find its way out of the Mediterranean and into the ocean.

**FIGURE 11.7**

### Outbreak of Typhoid in Mediterranean Resort
thought to be caused by drinking polluted water

Monk Seals, Marine Turtles and other Mediterranean wildlife threatened with extinction - conservationist's claim pollution from factories to blame

**Green algae growing in Adriatic Sea, causing sea-slime and sea-froth. Hotel owners claim fertilisers are to blame**

### Huge Oil Spill
as tanker runs aground near Greek coast. 6th incident this year along this coastline

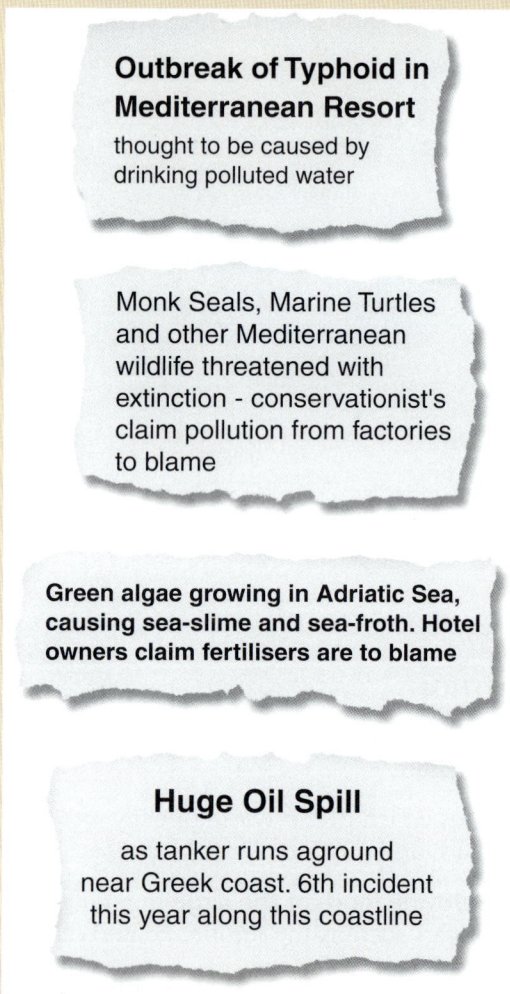

**FIGURE 11.8**   *Effects of pollution in the Mediterranean Sea*

## SOLUTIONS TO POLLUTION

In 1979 the countries around the Mediterranean agreed a plan to reduce pollution. It involved:

- Building more sewage treatment plants.

- Banning the dumping of some chemicals, for example mercury, DDT.

- Making factories have a licence before they are allowed to dump other chemicals.

- Marine reserves set up, in which farmers were paid not to use chemicals.

- Action to combat oil spillages (see Figure 11.10).

This plan has had some success. For example, in France the amount of mercury dumped in the Mediterranean has dropped by 90 per cent. No other plans since then have been successful because it has not been possible for all the countries to agree on ways of improving the environment.

**FIGURE 11.9**

## DEALING WITH OIL SPILLS

To deal with an oil spillage it is necessary to have people and equipment ready. Booms (floating barriers) are needed to trap the oil and prevent it from spreading. Then three methods can be used to get rid of the oil:

1 **Detergents** – these break up the oil but can also poison fish.

2 **Chalk** – this makes the oil sink, but it then affects plants and animals on the sea-bed

3 **Burning** – this only works if the oil is thick, but it also causes a lot of air pollution.

**FIGURE 11.10**

## OVERFISHING

There are hundreds of fishing ports along the coast of the Mediterranean where the fishermen catch a wide range of species, including sardines, tuna, mackerel and anchovy. But the numbers of fish have been falling recently and tuna, swordfish and sardines are now under threat.

**FIGURE 11.11**

## COMMON FISHERIES POLICY

One way of reducing overfishing would be to divide the Mediterranean into areas and only allow one country to fish in each area. But for this to work, all 21 countries would have to agree and this has so far proved impossible. The best plan has been the **Common Fisheries Policy**, but this only applies to those Mediterranean countries in the European Union.

The Common Fisheries Policy involves:

- Limiting the total catch of all fish species in any year – each country is given a quota (a maximum number of each species they are allowed to catch).

- Having a minimum mesh size for all fishing nets.

- No other country being allowed to fish within 19 km of a country's shoreline.

- Only European Union countries allowed to fish within 320 km of a member country's shore.

**FIGURE 11.12**

## 11G Sea and ocean pollution: problems and solutions

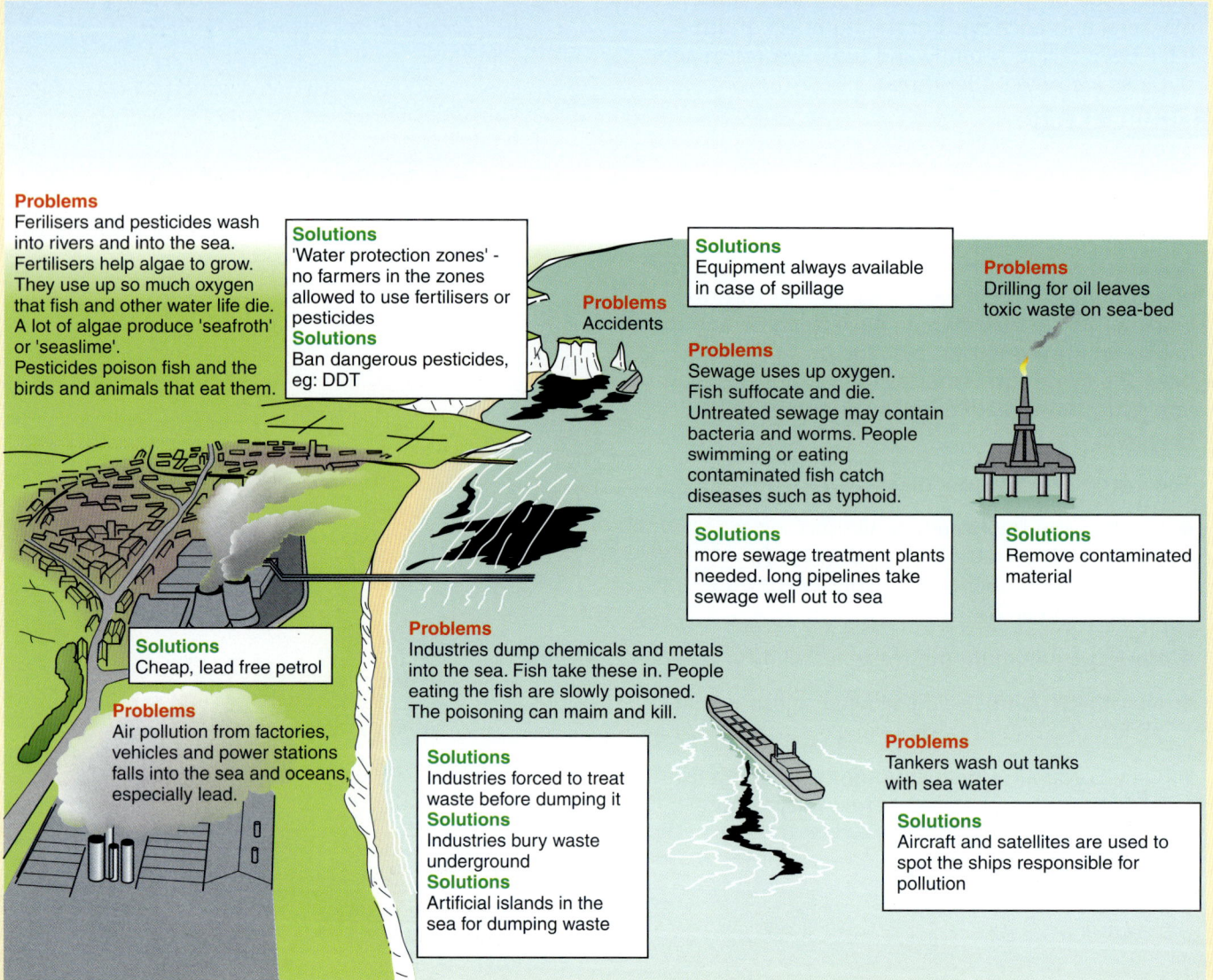

**Problems**
Feriliisers and pesticides wash into rivers and into the sea. Fertilisers help algae to grow. They use up so much oxygen that fish and other water life die. A lot of algae produce 'seafroth' or 'seaslime'.
Pesticides poison fish and the birds and animals that eat them.

**Solutions**
'Water protection zones' - no farmers in the zones allowed to use fertilisers or pesticides
**Solutions**
Ban dangerous pesticides, eg: DDT

**Problems**
Accidents

**Solutions**
Equipment always available in case of spillage

**Problems**
Drilling for oil leaves toxic waste on sea-bed

**Problems**
Sewage uses up oxygen. Fish suffocate and die. Untreated sewage may contain bacteria and worms. People swimming or eating contaminated fish catch diseases such as typhoid.

**Solutions**
more sewage treatment plants needed. long pipelines take sewage well out to sea

**Solutions**
Remove contaminated material

**Solutions**
Cheap, lead free petrol

**Problems**
Industries dump chemicals and metals into the sea. Fish take these in. People eating the fish are slowly poisoned. The poisoning can maim and kill.

**Problems**
Air pollution from factories, vehicles and power stations falls into the sea and oceans, especially lead.

**Solutions**
Industries forced to treat waste before dumping it
**Solutions**
Industries bury waste underground
**Solutions**
Artificial islands in the sea for dumping waste

**Problems**
Tankers wash out tanks with sea water

**Solutions**
Aircraft and satellites are used to spot the ships responsible for pollution

**FIGURE 11.13**

*Look at the Extension Text.*

1. Describe two causes of oil spills at sea.
2. Explain how fertilisers affect seas and oceans.
3. Explain how people can be poisoned by industrial waste.
4. Explain how sewage affects the oceans.
5. Describe how oil spills at sea can be reduced.
6. Describe how (a) industrial and (b) domestic pollution at sea can be reduced.

**FIGURE 11.14** *Beach pollution near Ipsos, Corfu, Greece*

## CREDIT QUESTIONS

### Case Study of the Mediterranean Sea

1. *Look at Figure 11.7.*
   Why does pollution remain in the Mediterranean Sea for a long time?

2. *Look at Figures 11.6 and 11.8.*
   Many tourist resorts are affected by the pollution. Which, do you think, is more serious to the tourist industry, the effects of dumping sewage or farm chemicals? Give reasons for your answer.

3. *Look at Figure 11.9.*
   Describe the different points of view towards marine reserves in the Mediterranean Sea.

4. *Look at Figure 11.10.*
   Which method of dealing with oil spills causes least environmental damage? Give reasons for your answer.

5. *Look at Figure 11.12.*
   Describe the arguments for and against countries being given fish quotas.

6. To what extent does the Common Fisheries Policy reduce overfishing?

# 12 Using the Land

## Core text

### 12A The world's food supply

The land is one of the world's most important resources. Every year it provides enough food to feed everyone in the world. But every year at least 40 million people die of hunger. Figure 12.1 shows how this happens.

Most of the people in the world live in developing countries, but they eat less than half of all the food. So it is in the developing world where most of the people suffering from hunger are found, and there are four main reasons for this:

1 Many farms are too small to make a living.
2 Many natural disasters ruin the farmland.
3 There is a lot of soil erosion and desertification.
4 Many people are in poor health.

### 12B Small farms

As the population in developing countries grows, so the farms become smaller. On small farms, farmers can only grow enough to feed themselves and their families (called **subsistence farming**). This means that they never have money to improve their land and grow more food.

### 12C Natural disasters

**Floods** wash away crops, animals, houses and even people.
**Cyclones** flatten crops, ruining harvests.
**Earthquakes** and **volcanoes** can also ruin farmland and kill people and animals.
**Pests** and **diseases** (for example locusts, rats, mice) eat one third of all the food grown.

A) WHERE DO ALL THE PEOPLE LIVE?

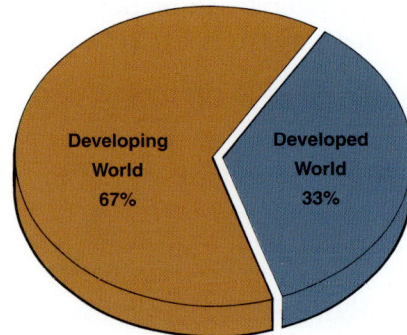

Developing World 67%

Developed World 33%

B) WHO EATS THE WORLD'S FOOD?

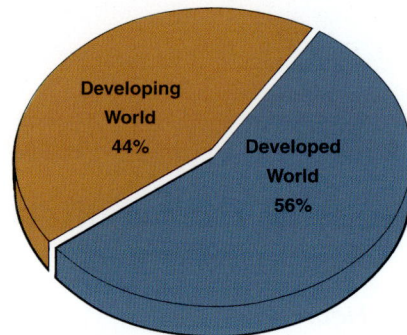

Developing World 44%

Developed World 56%

C) WHO PRODUCES THE WORLD'S FOOD?

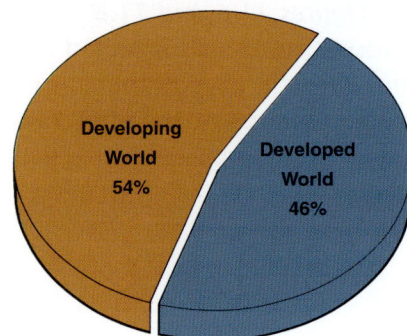

Developing World 54%

Developed World 46%

**FIGURE 12.1**

## 12D Poor health

Many people in the developing world suffer from disease, such as malaria, cholera, snail fever and sleeping sickness. Some of these are killer diseases, especially of children, but many more make people too weak to work. Once people are unhealthy, they become trapped in a vicious cycle of disease (see Figure 12.2).

## 12E Soil erosion and desertification

When soil is bare, it can easily be blown away by the wind or washed away by rain. This is called **soil erosion**. If a lot of soil is eroded the land cannot be farmed and it becomes desert. This is called **desertification**.

Desertification occurs when the soil becomes bare, for the following reasons:

1 **Drought** – with little rain, crops and grass do not grow and the soil turns to dust.
2 **Deforestation** (cutting down trees) – when trees are cut down, there are no roots to stop the soil from being eroded.
3 **Overpopulation** – with too many people living in an area, too many crops are grown on the land (**overcropping**) and too many animals are kept (**overgrazing**). The soil becomes so poor that little can grow and then it is easily eroded.

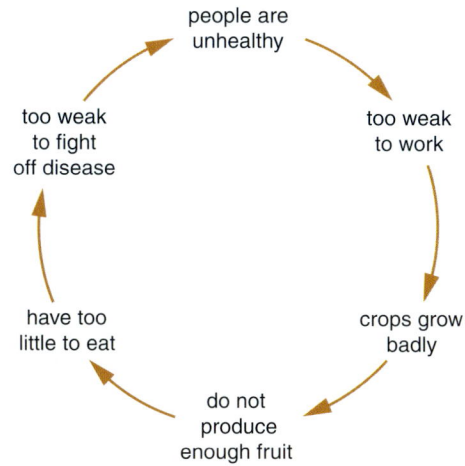

**FIGURE 12.2** *The disease cycle*

## 12F Methods of reducing desertification

| CAUSE | SOLUTION |
|---|---|
| **Drought** | **1 Irrigation** (putting extra water on farmland) from rivers, wells, reservoirs, so crops grow well and the land is not bare. |
| **Deforestation** | **1 Plant more trees** so their roots bind the soil together – the trees also make the soil more fertile. **2 Plant shelter belts** of trees to slow down the windspeed. |
| **Overpopulation** | **1 Use fertilisers** which enable crops to grow better. **2 Use improved crop varieties** developed by scientists, which have much higher yields. |

**FIGURE 12.3**

## 12G  The green revolution

Although many people in the developing world do not have enough to eat, Figure 12.4 shows that the situation is improving. Food production is increasing and it is increasing at a faster rate than the world's population. This great increase in food production in recent years is called the **green revolution**. It has been due to:

- Increased crop yields, by using new varieties and by using fertilisers and pesticides.

- Making more farmland for example by irrigating dry areas, terracing hillsides, draining marshes.

- Reducing soil erosion and desertification (see 12F).

- Sharing out farmland more equally.

| CALORIES PER PERSON PER DAY (WORLD AVERAGE) ||
|---|---|
| 1960 | 2334 calories |
| 1970 | 2488 calories |
| 1980 | 2600 calories |
| 1990 | 2697 calories |
| 2000 | 2807 calories |

**FIGURE 12.4**

## CORE QUESTIONS

1. *Look at Figure 12.1.*
   Choosing from: **developing countries** or **developed countries**,
   (a) which has more people?
   (b) which eats more of the world's food?
   (c) which produces more of the world's food?

2. *Look at 12B.*
   (a) What is subsistence farming?
   (b) Why do subsistence farmers find it difficult to grow more food?

3. *Look at 12C.*
   Name **three** natural disasters which can ruin farmland.

4. *Look at 12D.*
   If people in developing countries are ill, why are they also likely to go hungry?

5. *Look at 12E.*
   Name the **two** ways in which soil is eroded.

6. What is **desertification**?

7. Name **two** causes of desertification.

8. *Look at 12F.*
   In what ways do the following reduce desertification:
   (a) irrigation?
   (b) planting trees?

9. *Look at Figure 12.4.*
   Describe the changes in the amount of food eaten in the world since 1960.

10. *Look at 12G.*
    (a) What is the **green revolution**?
    (b) Describe **two** reasons for the green revolution.

## Case Study of the Sahel

**1** *Look at Figure 12.6.*
Where is the Sahel?

**2** *Look at Figure 12.7.*
How do people make a living in the Sahel?

**3** *Look at Figure 12.9.*
(a) What has happened to the amount of rainfall in the Sahel since 1968?
(b) How has this caused soil erosion and desertification?

**4** *Look at Figure 12.10.*
(a) Why do the people in the Sahel cut down so many trees?
(b) How does this cause soil erosion and desertification?

**5** *Look at Figure 12.13.*
Many parts of the Sahel suffer from desertification.
(a) How has this killed so many people?
(b) Why has this forced many people to move to cities?

**6** *Look at Figure 12.14.*
In what ways has the Gezira irrigation scheme in Sudan helped the local people?

**7** *Look at Figure 12.17.*
Do you think that planting trees in the Sahel is a good way of reducing desertification? Give reasons for your answer.

## GENERAL QUESTIONS

## Case Study of the Sahel

**1** *Look at Figures 12.5 and 12.6.*
Describe the location of the Sahel.

**2** *Look at Figure 12.7.*
How do the people of the Sahel cope with the low rainfall?

**3** *Look at Figure 12.8.*
Describe the changes in yearly rainfall in the Sahel since 1950.

**4** *Look at Figure 12.10.*
Describe how cutting down trees in the Sahel can lead to desertification.

**5** *Look at Figures 12.12 and 12.13.*
Describe the effects of desertification in the Sahel.

**6** *Look at Figures 12.14 and 12.17.*
Which is the better way of reducing desertification, irrigation schemes or planting trees? Give reasons for your answer.

**7** *Look at Figure 12.18.*
Describe one advantage and one disadvantage of **stone lines** as a way of reducing desertification.

# RESOURCES

## Case Study of the Sahel

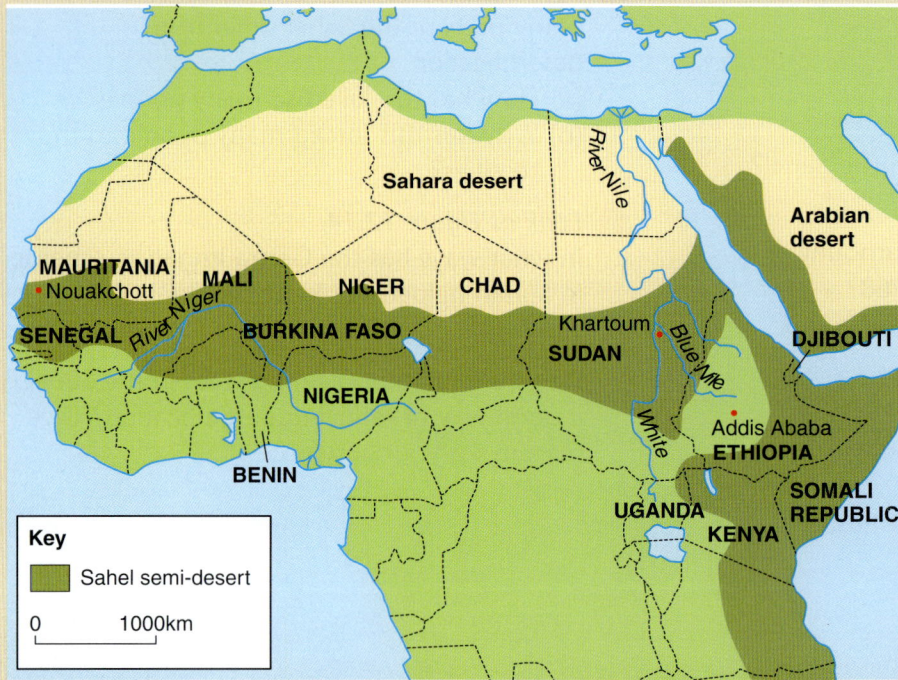

**FIGURE 12.5** The Sahel region

### INTRODUCTION

The amount of food being grown in the world is increasing more quickly than the population of the world. But this is not happening everywhere. In Africa, south of the Sahara Desert, there have been serious food shortages since 1968. This area is called the **Sahel** and the main reason for the food shortages is desertification – much of the land is turning into desert.

**FIGURE 12.6**

### TRADITIONAL WAY OF LIFE

The Sahel region is shown in Figure 12.5. It receives between 100mm and 500mm of rain each year. Even though it is very dry there is enough rain for grass to grow, so most people are herders of sheep, camels and goats. But, they need to keep moving from area to area to find enough water and grazing land (they are called **nomadic herders**). In the wetter parts, farmers can grow crops such as millet and sorghum. Over 40 million people live in this huge area which stretches more than 2000 kilometres across Africa.

**FIGURE 12.7**

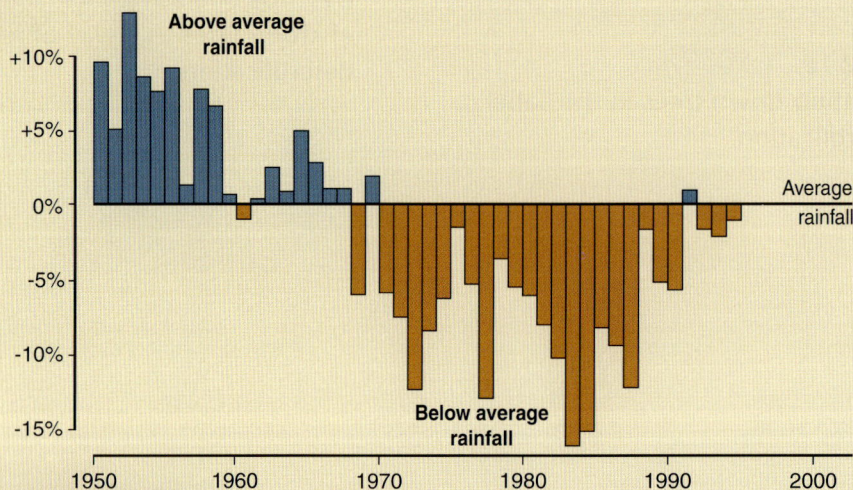

**FIGURE 12.8** Rainfall in the Sahel (1950–2000)

## PHYSICAL CAUSES OF DESERTIFICATION

Although the Sahel should get 100–500mm of rain each year, the rainfall is very unreliable. Some years are much wetter than average and some are much drier. The region can have many dry years in succession, and this has happened since 1968. With very little rain, few crops grew so there were few roots to hold the soil together. At the same time, the soil dried out and was easily blown away. The soil was eroded until some of the land began to turn into desert.

**FIGURE 12.9**

## HUMAN CAUSES OF DESERTIFICATION

People have made a living in the Sahel for hundreds of years without damaging the soil. But recently the population in this region has begun to grow rapidly. So people have been forced to grow crops on poorer land. When the crops fail, the bare soil is eroded by wind and rain. More people also means more animals are kept on the same land where they chew up all the grass or trample it and kill it. The result again is bare soil and soil erosion. Even the unfarmed areas suffer from soil erosion as the trees are cut down for firewood.

**FIGURE 12.11**

## EFFECTS OF DESERTIFICATION

In the last 30 years, desertification in the Sahel has brought extreme hardship and misery to the people of the Sahel and millions have died, either through starvatiion or diseases they were too weak to fight.

**FIGURE 12.12**

**Stage A**

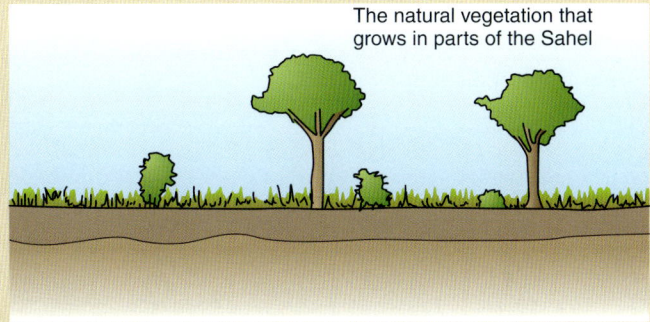

The natural vegetation that grows in parts of the Sahel

**Stage B**

Trees cut down for fuel and to make grazing land

**Stage C**

With no tree roots, the soil is blown away. When rains come, the soil is washed away, forming gullies

**Stage D**

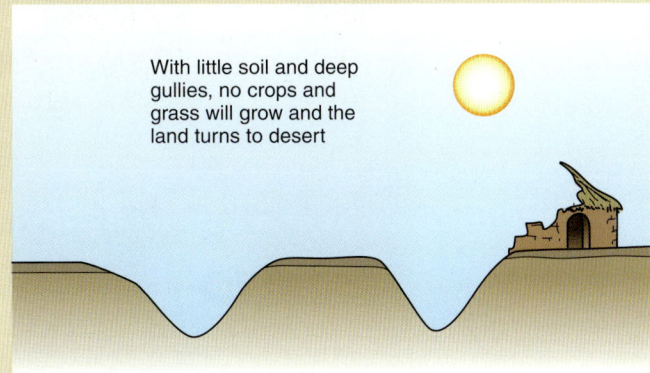

With little soil and deep gullies, no crops and grass will grow and the land turns to desert

**FIGURE 12.10** *Stages in the spread of the desert*

# RESOURCES

farmland slowly changes into desert

↓

crops and grass grow badly

↓

people go hungry

- people catch disease → people die
- people move to refugee camps or cities to try to survive there
- famine occurs → people starve

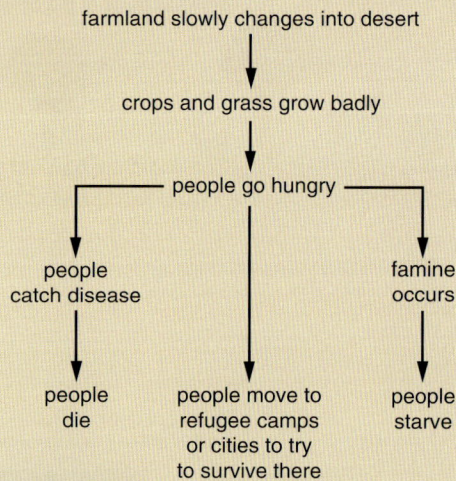

**FIGURE 12.13**  *Effects of desertification*

## SOLUTION 1: IRRIGATION

In the Sahel there are some expensive irrigation schemes where rivers have been dammed and reservoirs formed. The water from the reservoirs is then taken by canals to irrigate large areas of farmland.

In the Gezira scheme in the Sudan, one million hectares of land is irrigated using water from the White Nile and Blue Nile. This means that crops grow even if there is little rain. Farmers even have two harvests each year and their income has increased considerably. Cotton is grown for export, wheat and sorghum are grown for food and 150 000 people are now employed here. While it has brought many benefits, the Gezira scheme was expensive and only helped people in one small area of the country. The reservoir also flooded land which had previously been used by local farmers.

**FIGURE 12.14**

## SOLUTION 2: TERRACES

There are over 200 000 kilometres of terraces in Ethiopia. They help to trap water, stopping it from washing the soil away.

**FIGURE 12.15**

**FIGURE 12.16**  *Terracing in Ethiopia*

## SOLUTION 3: PLANTING TREES

Planting trees around villages in the Sahel stops sand dunes from spreading onto farmland and stops the soil being eroded. The trees provide shade for animals and they may supply products (for example honey, beans, gum) which can be sold. But they only do this when they are fully grown, and this takes many years. At El Ain in the Sudan, villagers have recently planted 200 000 tree seedlings. These trees will now be protected for sustainable use. They should provide enough fuel for the local population, slow down the speed of the wind and stop soil erosion taking place.

**FIGURE 12.17**

## SOLUTION 4: STONE LINES

In Burkina Faso and other Sahel countries local people have built lines of stones along the contours on their sloping farmland. These stone lines trap run-off after heavy rain so that the soil is not washed away. Crops also grow much better in the deeper soil behind the stones. This method requires a lot of labour, but it is cheap and simple to work.

**FIGURE 12.18**

**FIGURE 12.19** *Building stone lines in Burkina Faso*

## 12H Causes of desertification

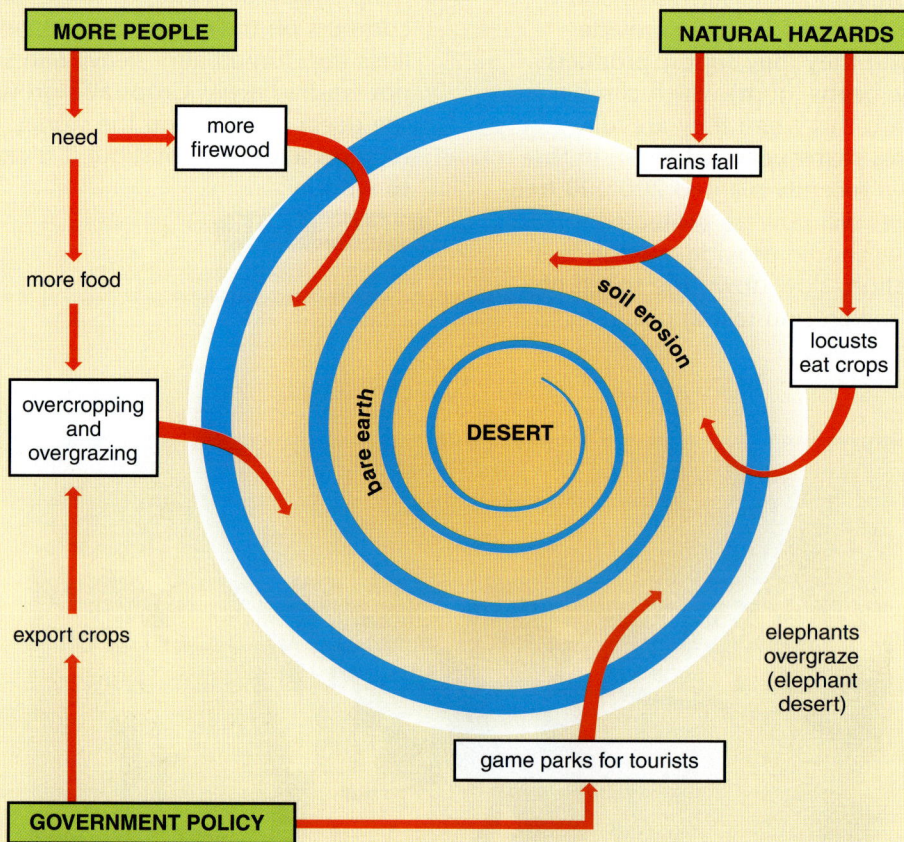

**FIGURE 12.20** Causes of desertification in Africa

Figure 12.20 shows the two main causes of desertification – population growth and reduced rainfall. Population growth has three consequences – it causes overgrazing, overcropping and deforestation.

Overgrazing occurs partly because farmers have less land and partly because they have more animals. In recent years farmers have been keeping more animals because their prices are so low.

Overcropping is partly due to people having less land but still needing to grow the same amount of crops and partly due to farmers now using poorer land which cannot be farmed so

intensively. One reason why farmers are using poorer land for growing food is that the best land is used to grow export crops.

Deforestation is taking place chiefly to provide fuel. In the developing world, wood is the main source of fuel. As the population grows, trees and bushes are cut down at a faster rate than they can regrow.

So, around settlements in the drier areas of the developing world, some land is being overgrazed, some is being overcropped and the land which is not farmed is losing its trees and bushes. The result is huge tracts of bare, loose soil which are rapidly turning into desert.

## >> EXTENSION QUESTIONS

*Look at 12H.*

1. In what way does the price of animals affect the amount of overgrazing?

2. Explain how growing export crops can increase desertification.

3. Explain why deforestation is taking place so rapidly in the developing world.

**Look at Figure 12.20.**

4. Suggest why the establishment of game parks may increase soil erosion.

## CREDIT QUESTIONS

### Case Study of the Sahel

1. *Look at Figure 12.7.*
   Explain how the people of the Sahel have adapted to the climate.

2. *Look at Figure 12.9.*
   Explain how the unreliable climate can lead to desertification.

3. *Look at Figures 12.10 and 12.11.*
   What is the relationship between population growth and desertification in the Sahel?

4. *Look at Figures 12.17 and 12.18.*
   Which is the better way of reducing desertification in the developing world, laying stone lines or planting trees? Give reasons for your answer.

5. *Look at Figure 12.14.*
   Describe the advantages and disadvantages of the Gezira irrigation scheme in the Sudan.

# 13 International Trade

## Core text

### 13A Introduction

We have already seen some of the ways in which countries try to improve their standard of living and how this creates a lot of environmental problems. In this unit we shall look at how countries try to develop by producing more goods and selling them abroad. This gives them money with which they can buy things which will raise the standard of living of their people. The buying and selling of goods between countries is called **international trade**.

### 13B Imports and exports

Goods which a country sells abroad are called **exports**. Goods which are bought from other countries are called **imports**. The difference between the value of a country's exports and its imports is called its **trade balance**.

Most developed countries receive more money from their exports than they have to pay for their imports. This is called a **trade surplus**. Most developing countries do not sell enough exports to pay for all the imports they need to buy. This is called a **trade deficit**.

Value of exports is greater than cost of imports = trade surplus.
Value of exports is less than cost of imports = trade deficit.

### 13C The pattern of international trade

**Manufactured goods** are those which have been made (for example steel, computers). **Primary goods** are foods and resources that have not been made but occur or grow naturally (for example coal, wheat, wood). Primary goods provide the raw materials for making manufactured goods. For example, wheat is needed to make bread, wood needed for making paper, coal needed for making steel.

Developed countries have many factories making manufactured goods, but they need primary goods as raw materials in their factories. So they import primary goods and export manufactured goods.

Developing countries have few factories so they need to import manufactured goods. They produce and export primary goods to pay for them. The table below shows other differences in trade between developed and developing countries.

|  | DEVELOPED COUNTRIES | DEVELOPING COUNTRIES |
|---|---|---|
| **Exports** | Mostly manufactured goods | Mostly primary goods |
| **Imports** | Primary and manufactured goods | Mostly manufactured goods |
| **Amount of trade** | Much | Little |
| **Number of exports** | Many | Very few |

**FIGURE 13.1** *International Trade*

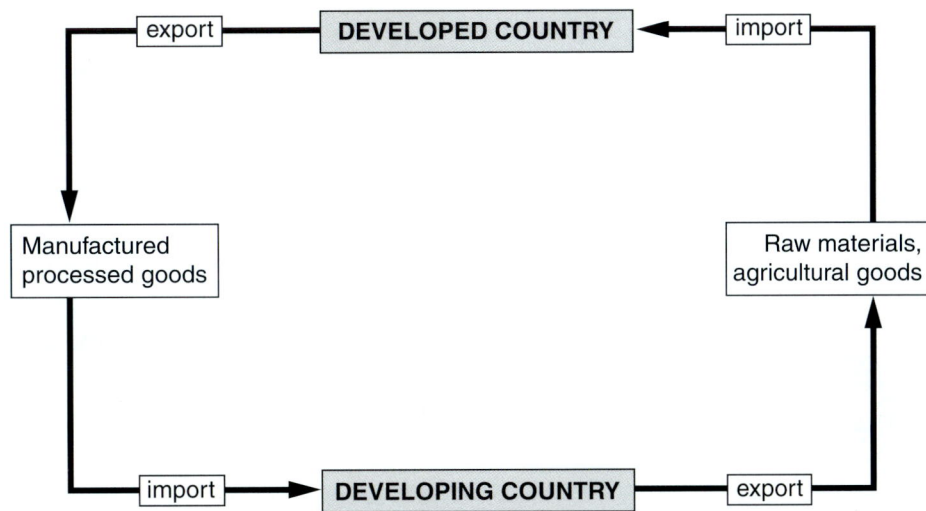

**FIGURE 13.2** *Interdependence between developed and developing countries*

## 13D Interdependence in trade

Developed countries need developing countries to supply them with primary goods which they use in their factories to make goods.

Developing countries need developed countries to supply them with the manufactured goods they cannot make themselves. So, developed and developing countries need each other and are said to be interdependent.

But it is still far from easy for developing countries to improve their standard of living by producing and exporting more goods. The rest of this unit explains why.

## 13E Problems in international trade

The cartoon in Figure 13.3 over the page shows some of the problems that developing countries face in international trade.

To sum up, primary goods are mostly cheap and go up and down in price a lot. Manufactured goods are mostly expensive and usually go up in price.

## 13F Barriers to international trade

Figure 13.4 shows some more problems facing developing countries as they try to trade. Developed countries make it difficult for other countries to export goods to them by:

1 Putting a **quota** or limit on the amount of goods they allow the country to export to them.
2 Adding **tariffs** or taxes to the price of imported goods, making them too expensive to buy.

FIGURE 13.3

1. *Look at 13B.*
   What are **imports** and **exports**?

2. What is a country's **trade balance**?

3. *Look at 13C.*
   For each of the goods below, write down whether they are manufactured or primary:
   (a) steel (b) wheat (c) machinery (d) tables (e) oil (f) fish.

4. How are the exports of developed and developing countries different?

5. *Look at 13D.*
   The cartoon in Figure 13.5 shows that developed and developing countries are interdependent. How does the cartoon show this?

6. *Look at 13E.*
   In what way do the prices of primary goods change?

7. How does this cause problems for developing countries?

8. In what way do the prices of manufactured goods change?

9. Why do developing countries need to borrow money?
   (a) 'It is a limit on the amount of goods a country can export to another country.'
   (b) 'It is a tax on imports.'

10. *Look at 13F.*
    Which of the statements above (A or B) describes (a) a tariff, and (b) a quota?

FIGURE 13.4 *Barriers to international trade*

## Case Study of Uganda

**1** *Look at Figure 13.9.*
(a) Since 1988 which have been greater, Uganda's exports or imports?
(b) Does this give Uganda any problems? Give a reason for your answer.

**2** *Look at Figure 13.10.*
(a) Does Uganda import mostly manufactured goods or primary goods?
(b) Does this give the country any problems? Give a reason for your answer.
(c) What percentage of Uganda's exports is coffee?
(d) Do you think it is good for Uganda to rely on one export? Give a reason for your answer.

**3** *Look at Figure 13.11.*
Why can Uganda now not afford the same amount of manufactured goods that it could buy in 1977?

**4** *Look at Figure 13.12.*
Since 1975, has the price of coffee (a) risen steadily, (b) fallen steadily, or (c) gone up and down a lot?

**5** *Look at Figure 13.13.*
When the price of coffee goes down, how does this affect coffee farmers in Uganda?

**6** *Look at Figures 13.14 and 13.16.*
Do you think people in Uganda should set up a factory making instant coffee powder? Give reasons for your answer.

**7** *Look at Figure 13.17.*
The Fairtrade Foundation tries to help farmers in the developing world.
What are the main ways it helps coffee growers in Uganda?

## Case Study of Uganda

**1** *Look at Figure 13.9.*
Describe Uganda's trade balance between 1988 and 1996.

**2** *Look at Figure 13.10.*
(a) Describe the characteristics of Uganda's exports and imports.
(b) In what ways do they cause Uganda problems?

**3** *Look at Figure 13.12.*
(a) Describe the trends in coffee prices between 1975 and 1997.
(b) Describe one advantage and one disadvantage to Uganda of the changes in coffee prices.

(c) Do you think the farmers are pleased with these changes in price? Give reasons for your answer.

**4** *Look at Figures 13.14 and 13.16.*
Give one argument for and one argument against Uganda starting up factories making instant coffee powder.

**5** *Look at Figure 13.17.*
The Fairtrade Foundation tries to help developing world farmers.
Which helps coffee growers in Uganda more, guaranteed prices or long-term contracts? Give reasons for your answer.

**FIGURE 13.5**

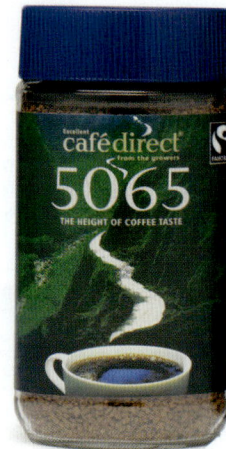

**FIGURE 13.6**

# Case Study of Uganda

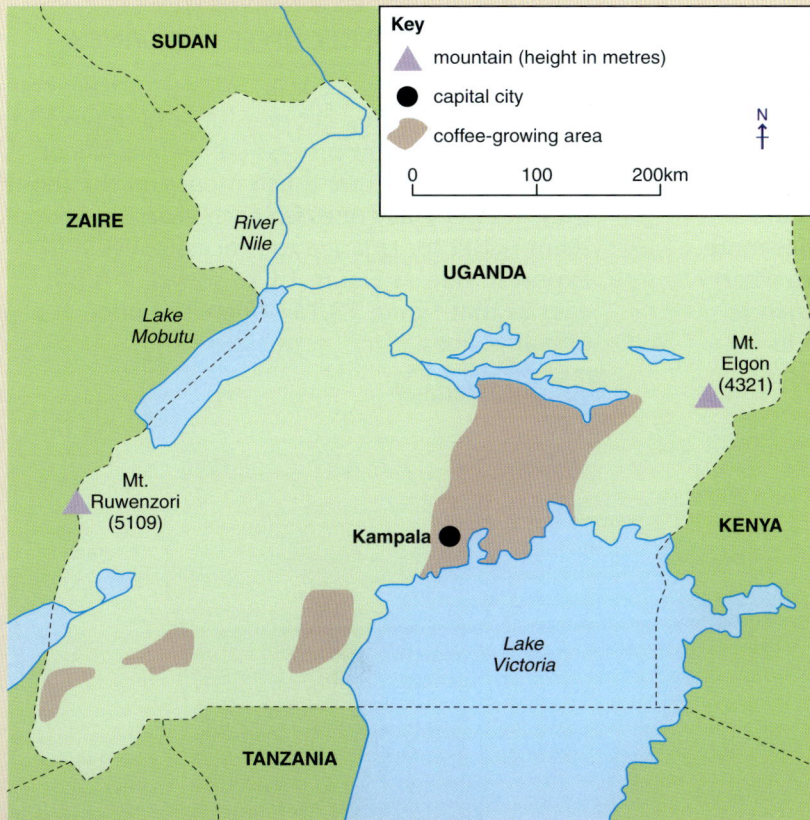

**Key**
- ▲ mountain (height in metres)
- ● capital city
- coffee-growing area

0    100    200km

N

SUDAN

ZAIRE

River Nile

Lake Mobutu

UGANDA

Mt. Elgon (4321)

Mt. Ruwenzori (5109)

Kampala

KENYA

Lake Victoria

TANZANIA

**FIGURE 13.8** *Uganda*

## INTRODUCTION

Uganda is a small country in east Africa. It is well above sea level, but the land is quite flat except on its borders where active volcanoes are found. Uganda has a tropical savanna climate well-suited to cattle farming, but a variety of crops are also grown. Despite this, the people of Uganda are among the poorest in the world. The government is trying to increase the country's wealth by producing more primary and manufactured goods and selling them abroad. But it faces many problems in doing this.

**FIGURE 13.7**

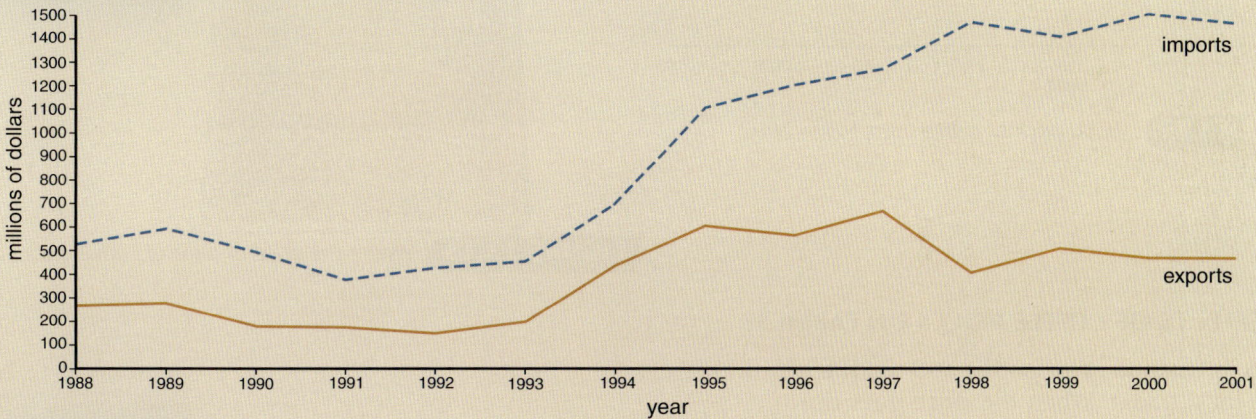

**FIGURE 13.9** *Line graph showing Uganda's imports and exports*

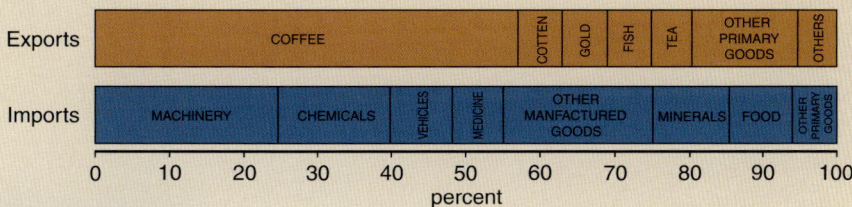

| Exports | COFFEE | COTTEN | GOLD | FISH | TEA | OTHER PRIMARY GOODS | OTHERS |
|---|---|---|---|---|---|---|---|

| Imports | MACHINERY | CHEMICALS | VEHICLES | MEDICINE | OTHER MANFACTURED GOODS | MINERALS | FOOD | OTHER PRIMARY GOODS |
|---|---|---|---|---|---|---|---|---|

0    10    20    30    40    50    60    70    80    90    100
percent

**FIGURE 13.10** *Breakdown of Uganda's imports and exports in 1999*

# RESOURCES

## THE EFFECTS OF COFFEE PRICES ON UGANDA

Uganda gets nearly three-quarters of its money from abroad by exporting coffee but, as Figure 13.12 shows, the price of coffee goes up and down a lot. When the price goes down, Uganda is hit very badly. For example in 1977, when the price was high, Uganda could buy five times more manufactured goods than it could in 1996, when the price was low. These are goods that Uganda needs to improve the standard of living of its people, such as farm machinery, fertiliser and hospital equipment.

**FIGURE 13.11**

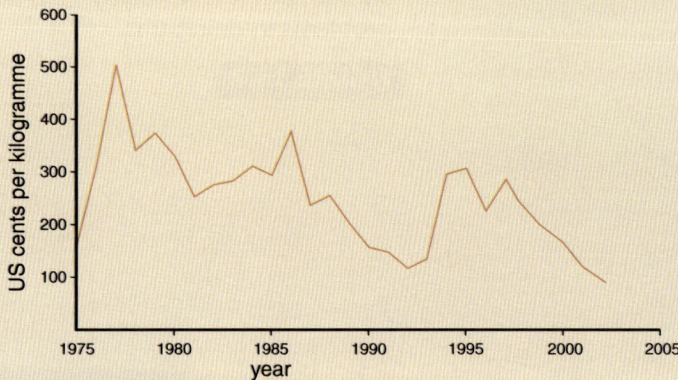

**FIGURE 13.12** *World price of coffee from 1975–2002*

## THE EFFECTS OF COFFEE PRICES ON FARMERS

Farmers in Uganda grow coffee and sell the beans to their local trader. When the price goes down they cannot afford to buy important goods such as food and clothes. They also find it very difficult to send their children to school and to go to hospital when they are ill, as these have to be paid for in Uganda. Nor is it easy for them to change to another type of farming. They would have to cut down all their coffee bushes and then buy different seeds and equipment. And, if they ever wanted to grow coffee again, the coffee bushes take five years to bear fruit.

**FIGURE 13.13**

## PROCESSING COFFEE IN UGANDA

Coffee growers receive little money for their coffee and the money they get goes up and down a lot. The companies that process coffee into jars of coffee powder receive much more and the money they get generally goes up each year. So, it would seem better for Uganda to process coffee and export it. But, as Figure 13.16 shows, it is not as easy as that.Figure 13.15 Where the price of a jar of coffee goes.

**FIGURE 13.14**

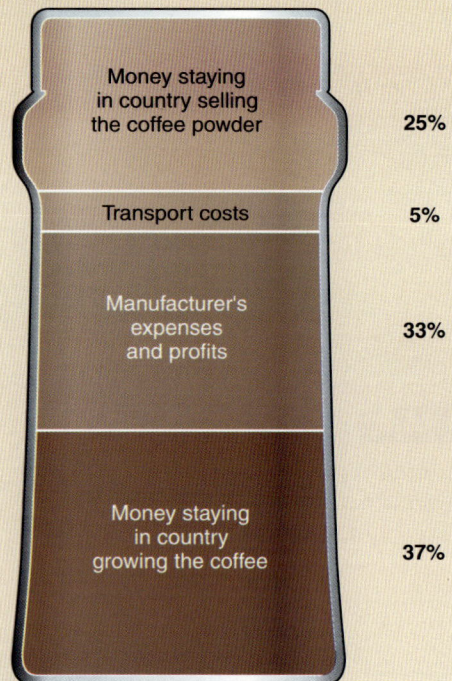

| | |
|---|---|
| Money staying in country selling the coffee powder | 25% |
| Transport costs | 5% |
| Manufacturer's expenses and profits | 33% |
| Money staying in country growing the coffee | 37% |

**FIGURE 13.15** *Where the price of a jar of coffee goes*

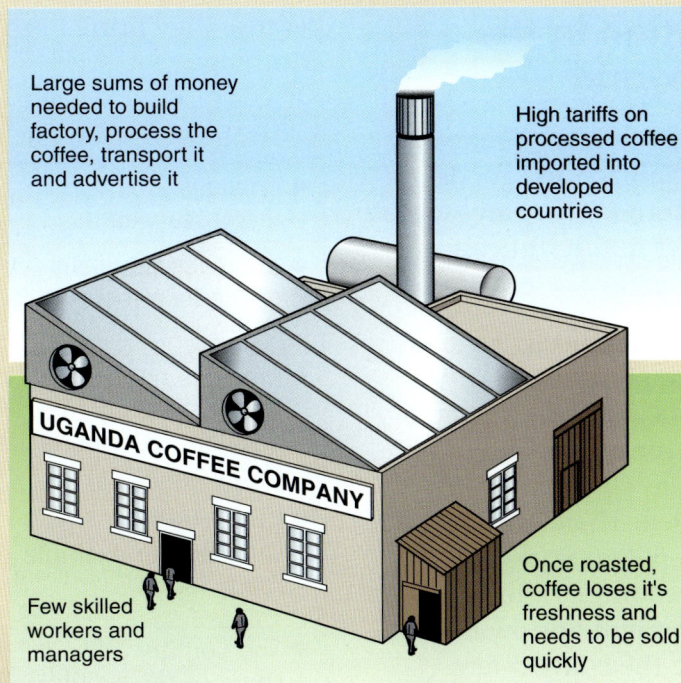

Large sums of money needed to build factory, process the coffee, transport it and advertise it

High tariffs on processed coffee imported into developed countries

UGANDA COFFEE COMPANY

Few skilled workers and managers

Once roasted, coffee loses it's freshness and needs to be sold quickly

**FIGURE 13.16** *Problems in setting up a coffee processing factory*

## THE FAIRTRADE FOUNDATION

Coffee growers in Uganda, and farmers all over the developing world, are being helped by organisations such as the Fairtrade Foundation. They check that companies making products from crops grown in the developing world offer:

1 Fair wages for their workers,

2 Fair terms of trade,

3 Decent working conditions,

4 Guaranteed prices, and

5 Long-term contracts.

Any company which reaches these standards is given the 'Fairtrade Mark', which is shown on their products. In this way, people in Britain can help developing world farmers by buying such products.

**FIGURE 13.17**

## 13G Industrialisation in the developing world

Developing countries export mostly primary goods. They export few manufactured goods because they have few factories. There are several reasons why these countries find it difficult to start up factories:

1 **Lack of capital** – any factory needs huge amounts of money before it produces any goods. The land has to be paid for, the factory has to be built and machinery and equipment installed. Services, such as electricity and water, need to be supplied. There are far fewer people with enough money to start up factories in developing countries and far fewer banks from which they might borrow money.

2 **Lack of skills** – because there are so few factories, there are few people skilled in working in and managing factories.

3 **Lack of a local market** – there are fewer people in developing countries who can afford to buy the goods made in any factory. This means that companies must sell abroad which makes their transport costs higher and their goods more expensive.

## 13H Multinational companies

One way for a developing country to overcome the problems involved in industrialisation is to use a **multinational** (or **transnational**) **company** to develop and export their goods. Multinational companies are companies with branches in many countries, but their headquarters are in a developed country, mostly the USA, for example General Motors, Pepsico, IBM. Their annual sales can be greater than the GNP of a developing country. For example, in 1994 the Mitsubishi company earned nearly $200 000 million. In the same year, the country of Ethiopia earned only $5000 million.

Multinational companies, however, may not be the answer to developing countries' problems because they bring both disadvantages as well as advantages.

| ADVANTAGES | DISADVANTAGES |
|---|---|
| ● Provides jobs and so increases wealth. | ● Jobs are often poorly paid |
| ● Improves the skills of the local people. | ● Many skilled workers are not local people. |
| ● Provides the money and technology to develop the country's resources. | ● Work is very mechanised and few employees are needed. |
| ● Improves the services, for example roads, ports. | ● Most of the profits go overseas. |
| ● The country takes a share of the profits. | ● They may pull out at any time. |

## >> EXTENSION QUESTIONS

*Look at the Extension Text.*

1 Why is there a lack of capital in developing countries to start up factories?

2 Why does the lack of a local market put off companies from setting up in a developing country?

3 What is a multinational company?

4 Do multinational companies increase the wealth of developing countries? Give reasons for your answer.

5 Is it true to say that multinational companies benefit developing countries by providing many jobs? Give reasons for your answer.

## Case Study of Uganda

**①** *Look at Figure 13.9.*
In what ways does Uganda's trade balance cause the country problems?

**②** *Look at Figure 13.10.*
Are Uganda's imports and exports typical of a developing country? Give reasons for your answer.

**③** *Look at Figures 13.11 and 13.12.*
Describe the problems to Uganda of relying so much on the export of coffee.

**④** *Look at Figures 13.15 and 13.16.*
Describe the different points of view towards setting up factories processing coffee in Uganda.

**⑤** *Look at Figure 13.13.*
Describe the problems faced by coffee farmers in Uganda.

**⑥** *Look at Figure 13.17.*
The Fairtrade Foundation offers benefits to coffee growers in Uganda. Which is of greater benefit, guarantee prices or long-term contracts? Give reasons for your answer.

# Core Groupwork

## The development game

This game is about a developing country trying to develop (See Figure 13.18).

TASK 1   You need a partner to play with and another couple to play against.

TASK 2   Agree how many turns to have each. One turn represents one year in the country.

TASK 3   Place the counter at the **start**. The start is where the people in the country have an average income of £2000 per year.

TASK 4   Each pair throws the dice in turn, moves and follows the instructions on the square it lands on.

TASK 5   While one partner throws the dice, the other writes down the information in a table (See table below).

TASK 6   When you have agreed to end the game, draw a line graph of your income over the years.

TASK 7   Write a report, explaining why your country's income has gone up and down.

| YEAR | AVERAGE INCOME | REASON FOR RISE OR FALL |
|---|---|---|
| 1 | | |
| 2 | | |
| 3 | | |
| 4 | | |
| 5 | | |

## 14D Why countries give aid

1. Aid helps trade. If aid helps developing countries to become richer, they will buy more goods from the developed countries.
2. Much official aid is **tied aid**. This means the country has to spend the money buying goods and services from the country that 'gave' it the money (called the **donor country**). So, this helps companies in the donor country.
3. Giving aid means the donor country has the support of the developing country, which it might need in times of war.
4. Many people in richer, developed countries believe they should help poorer countries.

## 14E Problems with giving aid

See Figure 14.1.

## 14F Self-help schemes

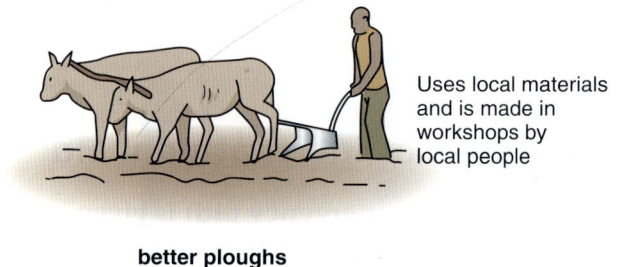

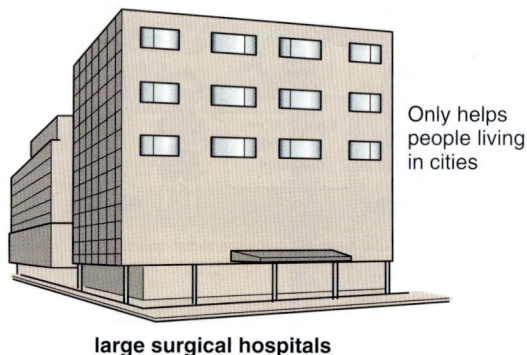

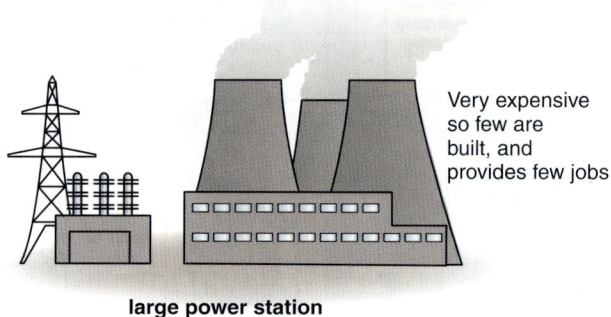

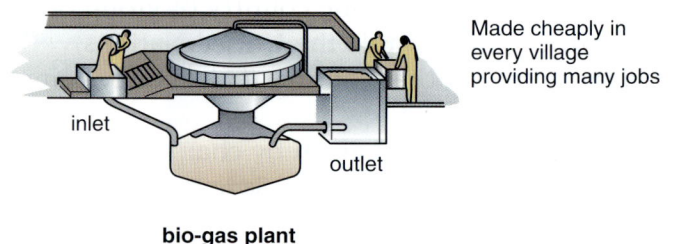

When international aid is used for large-scale projects, such as building dams and power stations, it does not usually benefit the people who need help most. An alternative is to use aid for **self-help schemes**. These schemes help people to help themselves, which means that they can continue to run the schemes after the aid has ended.

**INTERNATIONAL AID SCHEME**

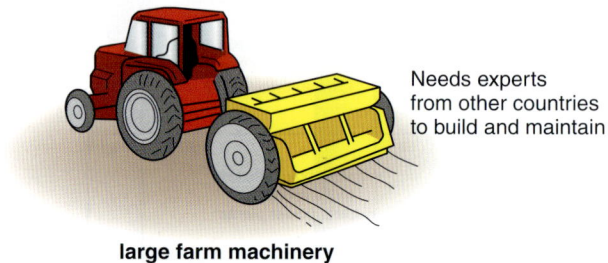

Needs experts from other countries to build and maintain

**large farm machinery**

Only helps people living in cities

**large surgical hospitals**

Very expensive so few are built, and provides few jobs

**large power station**

**SELF-HELP SCHEME**

Uses local materials and is made in workshops by local people

**better ploughs**

Local people trained as health workers, farming advisers

**village advisers**

Made cheaply in every village providing many jobs

inlet

outlet

**bio-gas plant**

**FIGURE 14.2** *International aid and self-help schemes*

Self-help schemes have the following characteristics:

1 They are **small-scale** – this makes them cheaper and means the country does not have to borrow money.
2 They **involve the local people** – this means they help everyone in the area and are more likely to be successful.
3 They **use simple technology** – so the skills of the local people are used and they do not have to depend on experts from other countries.

## CORE QUESTIONS

1 *Look at 14B.*
Give two reasons why a country might need short-term aid.

2 Name three examples of short-term aid.

3 Which of the following are examples of long-term aid:
(a) a new hospital,
(b) pumps to get rid of floodwater,
(c) new hotels for tourists,
(d) food and medicines?

4 *Look at 14C.*
What is 'official aid'?

5 Name two charities.

6 From where do charities get their money?

7 *Look at 14D and the cartoon below.*
What form of aid is the cartoon below trying to show?

8 Give two reasons why countries give aid to other countries.

9 *Look at 14E.*
Name two problems with giving aid.

10 *Look at 14F.*
Which of the following describe self-help schemes:
(a) they are big and expensive schemes
(b) they employ many local people
(c) they help all of the country
(d) they use advanced technology?

## FOUNDATION QUESTIONS

## Case Study of Ghana

1 *Look at Figure 14.6.*
(a) Which are greater, Ghana's imports or its exports?
(b) Does this give Ghana problems? Give reasons for your answer.

2 Are most people in Ghana healthy? Give reasons for your answer.

3 *Look at Figure 14.7.*
Describe the short-term aid that would have been needed after the war in 1994.

4 Many people died during the famine of 1983. Should all the aid money have been spent on giving people food or should some have been spent on improving farmland? Give reasons for your answer.

5 *Look at Figure 14.8.*
Do you think the Volta River Project was a sensible way of spending £100 million? Give reasons for your answer.

6 *Look at Figure 14.10.*
In western Ghana there is much disease. Which is the better way of solving this problem, scheme A or B in Figure 14.10? Give reasons for your answer.

7 *Look at Figure 14.11.*
Do you think the farming scheme in Figure 14.11 will help farmers to help themselves in the future? Give reasons for your answer.

8 *Look at Figure 14.12.*
Do you think new crop seeds will help all the farmers in Ghana? Give reasons for your answer.

## Case Study of Ghana

**1** *Look at Figure 14.6.*
In what ways does Ghana's trade balance cause problems for the country?

**2** In what ways does Ghana's natural increase cause problems for the country?

**3** *Look at Figure 14.7.*
Many people died during the famine of 1983. Should all the aid money have been spent on giving people food or should some have been spent on improving farmland? Give reasons for your answer.

**4** *Look at Figure 14.8.*
Describe two arguments for and two arguments against the Volta River Project..

**5** *Look at Figure 14.10.*
Which is the better way of solving the health problems in western Ghana, scheme A or scheme B in Figure 14.10? Give reasons for your answer.

**6** Do you think the scheme in Figure 14.11 will help farmers to help themselves in the future? Give reasons for your answer.

**7** Do you think the schemes in Figure 14.12 are self-help schemes? Give reasons for your answer.

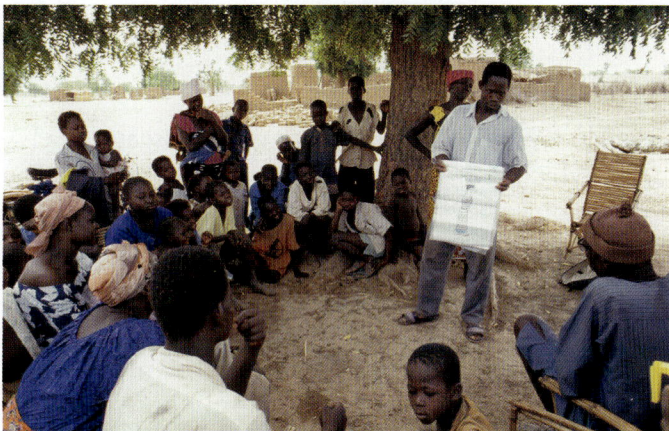

**FIGURE 14.3** *A health worker talking to villagers in Ghana*

## Case Study of Ghana

**1** *Look at Figure 14.6.*
Suggest reasons why Ghana needs international aid.

**2** *Look at Figure 14.8.*
Describe the different points of view people might have towards the Volta River Project.

**3** *Look at Figure 14.7.*
During the famine of 1983 describe the arguments for and against using all the aid available to help the starving people.

**4** *Look at Figure 14.10.*
Which is the better way of solving the health problems in western Ghana, scheme A or scheme B in Figure 14.10? Give reasons for your answer.

**5** *Look at Figure 14.11.*
Do you think that the farming project in Ghana is an example of a self-help scheme? Give reasons for your answer.

# RESOURCES

## Case Study of Ghana

### INTRODUCTION

Ghana is a developing country in west Africa. Most of its people are very poor. The population is rising rapidly and this is causing environmental problems, such as desertification in the north and deforestation in the south. The country only has two main exports, both of which are primary goods. Without help from other countries, Ghana would find it very difficult to improve the standard of living of its people.

**FIGURE 14.4**

### SHORT-TERM AID PROJECTS

Short-term aid is needed for emergencies such as those described below:

#### Daily News
*12th June 1994*

Fierce fighting in northern Ghana has left 6000 dead, 200 villages destroyed and 100 000 people homeless.

#### Daily Planet
*7th August 1983*

Severe drought and famine is affecting large areas of northern Ghana. Thousands of people are starving.

**FIGURE 14.7**

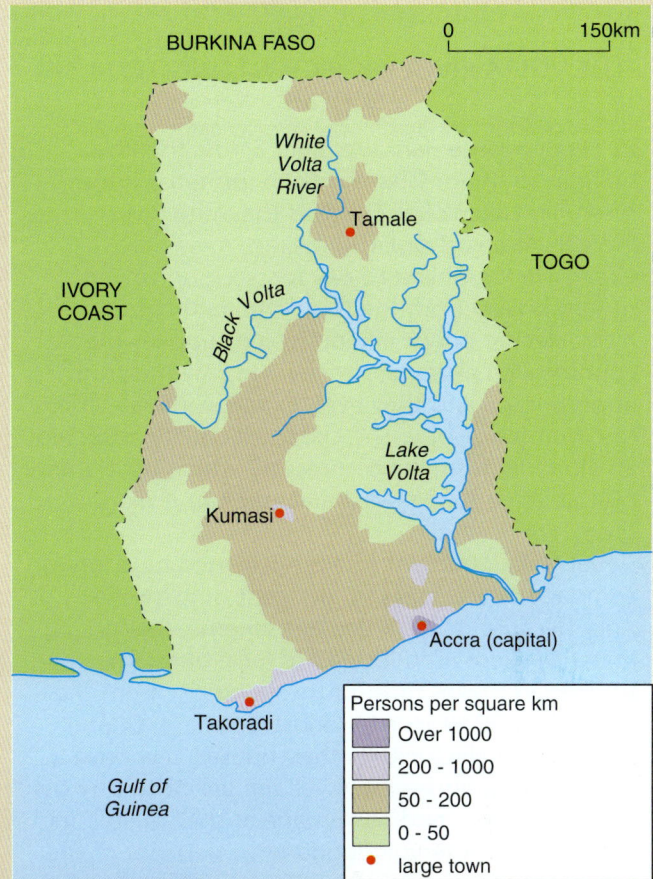

**FIGURE 14.5**  *The country of Ghana*

### FACTS ABOUT GHANA

| | |
|---|---|
| Population | 18 million |
| Population density | 80 people per sq km |
| Average income | £250 per year |
| Birth rate | 34‰ |
| Death rate | 11‰ |
| Imports | £670 million |
| Exports | £390 million |
| Main exports | cocoa (31%), gold (29%) |
| Average food eaten daily | 2200 calories |
| Food needed daily to be healthy | 2500 calories |
| Patients per doctor | 25 000 |
| Literacy rate | 64% |
| Official aid received | £440 million |

**FIGURE 14.6**

## THE VOLTA RIVER PROJECT: A LONG-TERM AID SCHEME

In 1966, a large dam was built across the River Volta in southern Ghana, which formed a huge reservoir, Lake Volta. This was the Volta River Project. At a cost of £100 million, it was too expensive for Ghana to pay for, so it borrowed money chiefly from the USA. The dam has a hydroelectric power station which supplies the country with cheap electricity. This saves it from importing oil and they even export some electricity to countries nearby. The cheap electricity attracted an aluminium smelting company which built a new port at Tema to export the aluminium it made there.

But the aluminium smelter is owned by a foreign company which takes 90 per cent of its profits. And, because it was a tied aid scheme, foreign materials were used instead of local ones. The smelter also takes most of the electricity from the dam. Over 30 years later, most people in Ghana still have no electricity in their houses, few have a water supply, and even fewer can afford to buy the aluminium pots and pans made at the smelter. In addition, the reservoir flooded five per cent of the whole country, forcing 80 000 people to move.

**FIGURE 14.8**

**FIGURE 14.9** *The Akosombo Dam on the river Volta, Ghana*

## A FARMING PROJECT IN CENTRAL GHANA

Central Ghana suffers from unreliable rain. Wells often run dry and so people can be short of water. Harvests are poor if the rains do not come.

A new farming project provides advice and materials for digging deeper wells. One person is trained in each village to maintain these wells.

**FIGURE 14.11**

## OTHER LONG-TERM PROJECTS

1  New crop breeds have been introduced which yield more food, but they need more fertiliser and pesticides to grow well.

2  Large power stations are being built at Takoradi and Tema`.

**FIGURE 14.12**

## HEALTH PROJECTS IN WESTERN GHANA

In western Ghana most people are subsistence farmers and many suffer from lack of food and malnutrition. Diseases are very common, such as malaria, guinea worm and diarrhoea.

| Scheme A (run by Christian Aid) | Scheme B (Official Aid Scheme) |
|---|---|
| 1  Advisers help people to lay stone lines and plant trees to improve the soil, so crops grow better | 1  Large hospital, fully equipped and staffed, to be built in nearest city. |
| 2  They build new clinics and train local people to recognise and prevent the most common diseases. | |

**FIGURE 14.10**

## 14G Categories of aid

International aid comes in a variety of forms. For example, aid can be:

- **Technical aid** – advisers, consultants, specialists
- **Project aid** – for specific projects only, for example a power station
- **Emergency aid** – food, clothing, shelter, medicines
- **Tied aid** – money which has to be used to buy goods from the donor country
- **Grants** – money given with no 'strings' attached
- **Loans** – money lent which has to be repaid with interest

Aid also uses different forms of technology. Some involve **advanced technology**, such as building dams, motorways. These often require skills that the local people do not have. Other types of aid, especially self-help schemes, involve **intermediate technology**. This uses simpler equipment that the local people can be trained to use. It also costs less. For these reasons, it is also called **appropriate technology**.

Aid also comes from a variety of sources, which all have their advantages and disadvantages, as shown below.

## >> EXTENSION QUESTIONS

*Look at the Extension Text.*

1. What is the difference between bilateral and multilateral aid?

2. Which source of aid:
   (a) most uses advanced technology schemes,
   (b) usually involves self-help schemes,
   (c) is often tied aid,
   (d) gives the most money,
   (e) lends much money?

3. Explain why intermediate technology is often called appropriate technology.

| SOURCE OF AID | ADVANTAGES | DISADVANTAGES |
|---|---|---|
| **Bilateral Aid** (from one country to another) | 1 Quick and direct.<br>2 Has large funds for major schemes. | 1 Mostly 'tied aid'.<br>2 Receiving country is tied politically to donor country.<br>3 Often used for large prestige projects using advanced technology. |
| **Multilateral Aid** (from several countries) | 1 Not 'tied'.<br>2 No political strings. | 1 Mostly loans, which have to be repaid with interest.<br>2 Takes longer for the aid to be given. |
| **Voluntary Aid** (from charities) | 1 No political strings.<br>2 Encourage self-help schemes, using appropriate technology. | 1 Very limited funds.<br>2 Funds are less reliable, as they depend on gifts and donations. |

# 15 International Influence and Alliances

## Core text

### 15A International influence

If a country is going to develop, it needs help from other countries. It needs them to trade with, to provide them with aid or even to provide them with defence.

Countries which can provide a lot of trade, aid and defence to others have a lot of **international influence**. They can influence the way other countries are run.

### 15B Measuring international influence

A country's international influence depends upon five main factors:

**1 The size of the Country**
Large countries can have a lot of influence. They control large areas of land and all the resources on it and underneath it. They do not need to depend as much on other countries. But not all large countries have a lot of influence. Some have very few people and few resources.

**2 The population of the country**
Countries with a large population can have a lot of influence. A large population can produce a lot of wealth and the country can have a large number of armed forces. But not all countries with a large population have a lot of influence. In some countries the people are not healthy enough or educated enough to develop the country fully.

**3 The resources of the country**
Countries with a lot of resources can have a lot of influence. The resources can be used to make the country wealthy. But some countries are too poor to use all the resources they have. And some resources are more important than others. Coal, iron ore and oil are valuable because so many industries need them.

**4 The Military strength of a country**
Countries with a lot of arms and armed forces have a lot of influence. They can provide arms or military advice to other countries. They can even attack other countries.

**5 The wealth of a country**
Countries with a large amount of wealth can have a lot of influence. They can afford to buy goods from other countries and provide a lot of aid. They are also rich enough to build up huge military strength.

| TOP 5 COUNTRIES (AREA) |
|---|
| 1 Russia |
| 2 Canada |
| 3 China |
| 4 USA |
| 5 Brazil |

| TOP 5 COUNTRIES (POPULATION) |
|---|
| 1 China |
| 2 India |
| 3 USA |
| 4 Indonesia |
| 5 Brazil |

| TOP 5 COUNTRIES (COAL, OIL, IRON ORE) |
|---|
| 1 China |
| 2 USA |
| 3 Russia |
| 4 Saudi Arabia |
| 5 India |

| TOP 5 COUNTRIES (ARMED FORCES) |
| --- |
| 1 China |
| 2 USA |
| 3 Russia |
| 4 India |
| 5 Pakistan |

| TOP 5 COUNTRIES (WEALTH) |
| --- |
| 1 USA |
| 2 Japan |
| 3 Germany |
| 4 France |
| 5 UK |

## 15C Alliances

Because countries need each other for aid, for trade or for defence, many have grouped themselves together. Groups of countries that work together are called **alliances**.

## 15D Trade or economic alliances

Some countries form alliances to make it easier to trade with each other. Each country in a trade alliance is allowed to sell its goods in all the other member countries. But if outside countries want to sell their goods, a **tariff** (a tax on the price) or a **quota** (a limit) is put on their goods. Examples of trade alliances include the European Union and the North American Free Trade Agreement (NAFTA).

## 15E Selling alliances

Countries which sell the same goods can get a higher price if they form an alliance and agree to sell at the same price. An example of a selling alliance is OPEC (Organisation of Petroleum Exporting Countries). These countries (which include Saudi Arabia, Iran, Kuwait) formed an alliance to sell oil at a high price. For example, in 1973 they raised the price by 400 per cent. Other countries had to pay up because they could not get their oil from anywhere else.

## CORE QUESTIONS

1 *Look at 15B.*
 Name the five ways of measuring a country's influence.

2 Explain why some large countries have little influence.

3 Explain why some countries with a large population have little influence.

4 *Look at 15C.*
 What is an alliance?

5 *Look at 15D.*
 Why do countries form trade alliances?

6 What is (a) a tariff, and (b) a quota?

7 Give one example of a trade alliance.

8 *Look at 15E.*
 Why do countries form selling alliances?

9 Name one selling alliance.

**FIGURE 15.1** *How a trade alliance works*

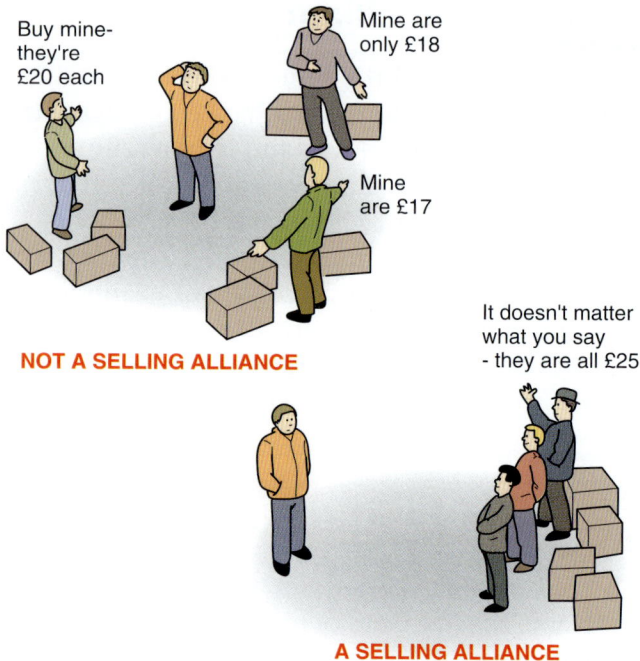

Buy mine-
they're
£20 each

Mine are
only £18

Mine
are £17

NOT A SELLING ALLIANCE

It doesn't matter
what you say
- they are all £25

A SELLING ALLIANCE

**FIGURE 15.2** *How a selling alliance works*

## FOUNDATION QUESTIONS

### Case Study of the European Union

**1** *Look at Figure 15.5.*
What type of alliance is the European Union?

**2** Is the European Union getting larger or smaller?

**3** What is meant by the EU's single market?

**4** *Look at Figure 15.7.*
(a) Why is it difficult for other countries to sell their goods in the EU?
(b) How does this affect companies in the EU?

**5** The EU has removed border controls.
How does this affect industries?

**6** *Look at Figure 15.8.*
Goods from other countries become more expensive when they are sold in the EU.
Which becomes more expensive,
(a) Coffee beans or coffee powder?
(b) Cocoa beans or chocolate?

**7** *Look at Figure 15.10.*
Do you think farmers in the UK are pleased that there is free trade within the EU countries? Give reasons for your answers.

**8** *Look at Figure 15.12.*
Choosing from the EU or USA
(a) Which is larger?
(b) Which has more people?
(c) Which has more wealth?
(d) Which has more international influence?
Give reasons for your answer.

## Case Study of the European Union

1 *Look at Figure 15.5.*
Describe how the number of countries in the European Union has changed?

2 Describe what is meant by the EU's single market.

3 *Look at Figure 15.7.*
Do you think UK companies are pleased there is free trade between the EU countries? Give reasons for your answer.

4 In what ways are our industries protected by belonging to the EU?

5 *Look at Figure 15.8.*
Compare the tariffs on manufactured goods and primary goods coming into the EU.

6 *Look at Figure 15.12.*
Which has the most international influence, the EU, USA or Japan? Give reasons for your answer.

7 Do you think Japan is a superpower, with a lot of influence? Give reasons for your answer.

**FIGURE 15.3** *The headquarters of the European Union in Brussels*

# Case Study of the European Union

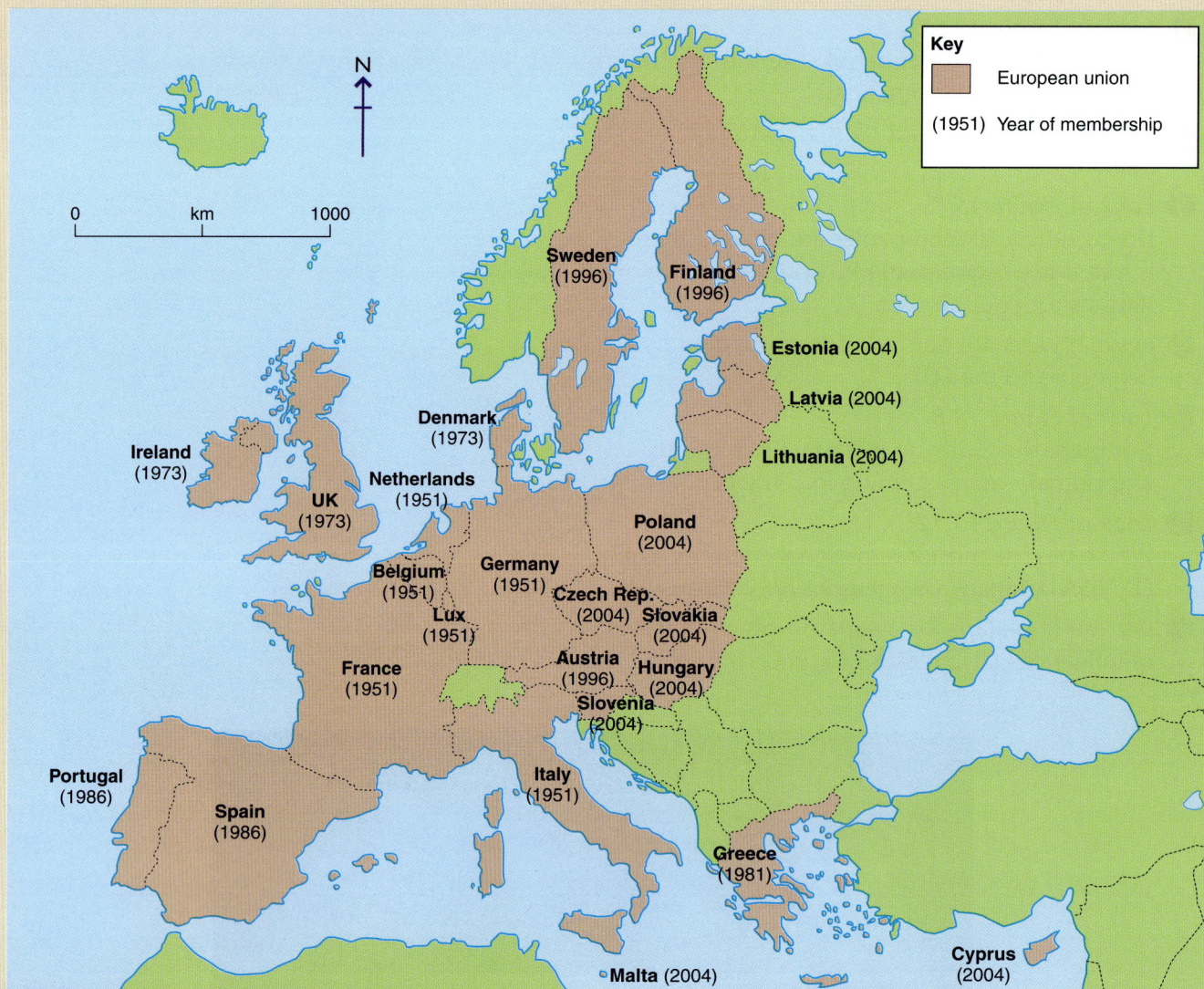

**FIGURE 15.4** The countries of the European Union

## INTRODUCTION

The European Union (EU) began as a small trade alliance of six countries in 1951. It increased to nine when the UK, Denmark and Ireland joined in 1973. By 1996 15 countries belonged and this increased to 25 countries in 2004 (see Figure 15.4).

Like all alliances, these countries have joined together to get benefits and one of the main benefits of belonging to the EU is that there is a **single market**. This means there is free trade and free movement of people between all the member countries. This affects everybody who lives in the EU, but in particular it affects farming and industry.

**FIGURE 15.5**

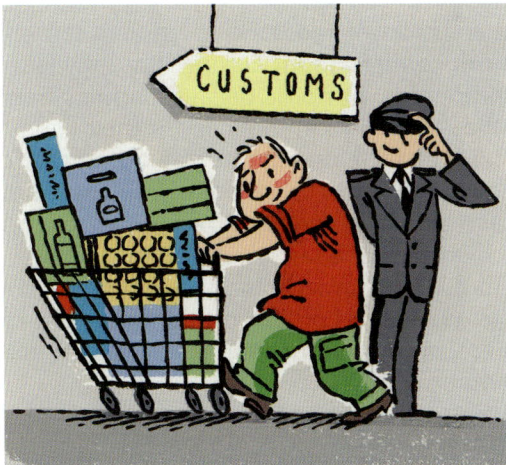

The EU will abolish customs duties. This will make goods cheaper.

The EU makes it easier for people to work in different parts of the European Union.

**FIGURE 15.6** *The effects of the EU on people*

## THE EFFECTS OF THE EU ON INDUSTRY

1 The EU allows **free trade** between member countries (no tariffs or quotas are allowed). This makes it easier for companies to sell their goods to other member countries. But, it also means more competition at home because companies from other EU countries can now sell their products there.

2 The EU has set up **trade barriers** (tariffs and quotas) with the rest of the world. This makes it more difficult for other countries to sell their goods in Europe. This means our companies are protected and will sell more and so make more profit and employ more people.

3 The EU has **removed border controls** (for example customs checks). This makes it quicker and cheaper to transport goods throughout Europe.

4 The EU also **gives money** to attract industries to areas of Europe which have a lower standard of living and where unemployment is very high.

**FIGURE 15.7**

| IMPORT | TARIFF |
|---|---|
| Cashmere jumpers | 13% |
| Chocolate | 10% |
| Cocoa beans | 1% |
| Coffee beans | 0% |
| Coffee powder | 14% |
| Copper | 0% |
| Flour | 15% |
| Iron ore | 0% |
| Tin | 0% |
| Video recorders | 14% |

**FIGURE 15.8** *Examples of EU Tariffs*

**FIGURE 15.9** *Competition from EU Farmers*

## THE EFFECTS OF THE EU ON FARMING

1  The EU allows free trade so farmers can sell their produce throughout Europe. But they face more competition at home because farmers from other countries can now sell in their country.

2  The EU has helped farmers by giving them **guaranteed prices** for their products. These have been high enough for farmers to earn a good living. If a farmer could not sell his products to anyone else, the EU bought them at the guaranteed price. This has been part of the EU's Common Agricultural Policy. It has made farming less risky.

**FIGURE 15.10**

## THE EUROPEAN UNION'S INFLUENCE

As more and more countries have joined the EU, so its power and influence has grown. Along with the USA and Japan, the EU is now one of the world's three superpowers and its international influence is shown in Figure 15.12.

**FIGURE 15.11**

|  | JAPAN | EUROPEAN UNION | USA |
|---|---|---|---|
| **Area** (thousand sq km) | 400 | 3300 | 9800 |
| **Population** (millions) | 130 | 370 | 270 |
| **Military Strength** (thousands of forces) | 200 | 2300 | 1500 |
| **Wealth** (million dollars) | 5000 | 8000 | 7000 |
| **Resources** (world position) | 3rd for eggs<br>4th for fish | 1st for barley, milk<br>2nd for potatoes,<br>wheat | 1st for maize,<br>2nd for milk, pigs,<br>coal, copper, gold,<br>oil and gas |

**FIGURE 15.12**

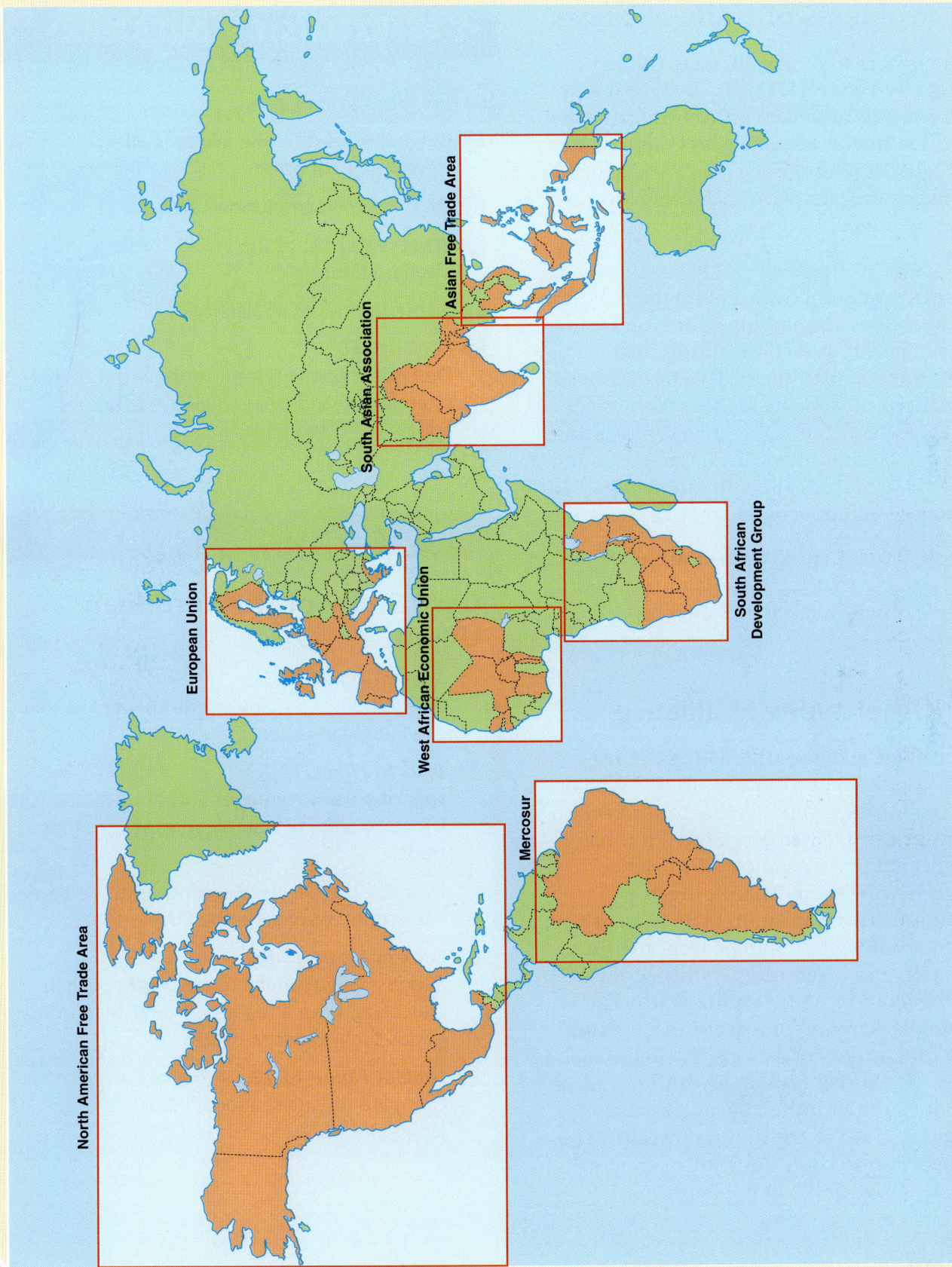

**FIGURE 15.13** *Major world trade alliances*

North American Free Trade Area

European Union

West African Economic Union

South Asian Association

Asian Free Trade Area

South African Development Group

Mercosur

## 15F  Problems of trade alliances

There are now many trade alliances throughout the world (see Figure 15.13). They make it easy for each country to trade within the alliance but much more difficult to trade with countries in another alliance. Developing countries, which are often in alliances only with each other, find it increasingly difficult to export to the richer countries.

The countries of Europe are among the richest in the world and, by forming the European Union, they have made it more difficult for other countries to trade with them. This has deprived poorer nations of valuable markets. Because of this an agreement, called the Lomé Convention, between the EU and the African, Caribbean and Pacific (ACP) countries allows them to export to the EU agreed amounts of goods at agreed prices.

In 1995 the **World Trade Organisation** was set up to ensure fair trade between countries and to settle trade disputes. It has 125 member countries.

## 15G  Other types of alliance

As well as trade alliances and selling alliances, there are also **defence alliances**. In this type of alliance, countries group together for protection. A country is less likely to be attacked if it has allies to defend it. An example of a defence alliance is NATO (North Atlantic Treaty Organisation). This was set up in 1949 to protect the countries of western Europe from the then, Soviet Union and communist countries of eastern Europe. In addition, some countries which share a common background or interest may group together for a variety of reasons. These are **social alliances**. The British Commonwealth is a social alliance of countries united by an allegiance to the Queen or Head of the Commonwealth. The Arab League unites countries with a common religion, language and cultural heritage.

## >> EXTENSION QUESTIONS

*Look at the Extension Text.*

1. In what ways do trade alliances affect developing countries?

2. Describe the Lomé Convention.

3. What is the role of the World Trade Organisation?

4. Name four types of alliance.

5. *Look at 15B.*
Describe a points system to show the overall international influence of all the countries appearing in the tables in 15B.

## CREDIT QUESTIONS

### Case Study of the European Union

1. *Look at Figure 15.5.*
Suggest why so many countries want to join the European Union.

2. *Look at Figure 15.7.*
Describe the advantages and disadvantages to UK companies of a single market with free trade.

3. Describe the effects of setting up trade barriers with countries outside the EU.

4. *Look at Figure 15.8.*
Suggest why manufactured goods, such as chocolate, have a higher EU tariff than primary goods such as cocoa beans.

5. *Look at Figure 15.12.*
Compare the international influence of the EU, Japan and USA.

# Index